REA: THE TEST PREP AP® TEACHERS RECOMMEND

AP® ENVIRONMENTAL SCIENCE
ALL ACCESS®

Kevin R. Reel
Head of School
The Colorado Springs School
Colorado Springs, Colorado

Research & Education Association
Visit our website: www.rea.com

Research & Education Association
61 Ethel Road West
Piscataway, New Jersey 08854
E-mail: info@rea.com

AP® ENVIRONMENTAL SCIENCE ALL ACCESS®

Published 2015

Copyright © 2014 by Research & Education Association, Inc. All rights reserved. No part of this book may be reproduced in any form without permission of the publisher.

Printed in the United States of America

Library of Congress Control Number 2013941784

ISBN-13: 978-0-7386-1082-5
ISBN-10: 0-7386-1082-8

LIMIT OF LIABILITY/DISCLAIMER OF WARRANTY: Publication of this work is for the purpose of test preparation and related use and subjects as set forth herein. While every effort has been made to achieve a work of high quality, neither Research & Education Association, Inc., nor the authors and other contributors of this work guarantee the accuracy or completeness of or assume any liability in connection with the information and opinions contained herein and in REA's software and/or online materials. REA and the authors and other contributors shall in no event be liable for any personal injury, property or other damages of any nature whatsoever, whether special, indirect, consequential or compensatory, directly or indirectly resulting from the publication, use or reliance upon this work.

AP® is a registered trademark of the College Board, which was not involved in the production of, and does not endorse, this product. All other trademarks cited in this publication are the property of their respective owners.

Cover image: © iStockphoto.com/damircudic

 All Access® and REA® are registered trademarks of Research & Education Association, Inc.

Contents

About Our Author vii
About Research & Education Association vii
Acknowledgments vii

Chapter 1: Welcome to REA's All Access for AP Environmental Science   1

Chapter 2: Strategies for the Exam   7

Chapter 3: Basic Concepts and Skills Review   23

Review of Principles of Matter 23
Review of Math Skills 25
Analyzing Scientific Data 28
Review of Experimental Design 32

Chapter 4: Earth Systems   35

Global Movement of Matter and Energy 36
The Earth 36
The Atmosphere 40
Global Water Resources 47
Soil Resources 51
Quiz 1 .... *available online at www.rea.com/studycenter*

Chapter 5: The Living World — 57

Ecosystem Structure 58

Survey of Earth's Ecosystems 61

Energy Flow in Ecosystems 65

Ecosystem Diversity and Change 67

Natural Biogeochemical Cycles 70

Quiz 2 .... *available online at www.rea.com/studycenter*

Chapter 6: Populations — 75

Population Biology 76

Human Populations 84

Quiz 3 .... *available online at www.rea.com/studycenter*

Mini-Test 1 .... *available online at www.rea.com/studycenter*

Chapter 7: Land and Water Use — 93

Agriculture 94

The Environmental Impact of Agriculture 99

Pest Control 100

Rangelands 101

Fishing 102

Forests 104

Wetlands 105

Mining 107

Urban Land Development 109

Transportation Infrastructure 111

Public and Federal Lands 113

Land Conservation Options... 115

Water Resources Management 116

Global Economics... 121

Quiz 4..*available online at www.rea.com/studycenter*

Chapter 8: Energy Resources and Consumption — 127

Energy Concepts... 128

Energy Units and Calculations.. 129

Human Energy Consumption... 132

Nonrenewable Energy Sources 134

Renewable Energy Sources.. 143

Energy Conservation.. 147

Quiz 5..*available online at www.rea.com/studycenter*

Chapter 9: Pollution — 151

Air Pollution 152

Effects of Pollutants on Humans 161

Effects of Pollutants on Ecosystems.. 162

Noise Pollution 165

Water Pollution.. 166

Solid Waste 173

Effects of Pollution on the Environment and Humans.................................... 177

LEED Certification.. 185

Quiz 6..*available online at www.rea.com/studycenter*

Chapter 10: Global Changes — 189

Stratospheric Ozone 190
Global Warming 192
Biodiversity 203
Quiz 7 .... *available online at www.rea.com/studycenter*
Mini-Test 2 .... *available online at www.rea.com/studycenter*

Practice Exam (also available online at www.rea.com/studycenter) — 211

Answer Key 240
Detailed Explanations of Answers 241

Glossary — 259

Index — 271

About Our Author

Kevin R. Reel has been involved in science education and school leadership for 30 years. During that time he has written numerous articles for academic journals and contributed to several textbooks in chemistry, health, and environmental science. He has served as the science department head at Thacher School, high school principal at The Westminster Schools, and headmaster at The Colorado Springs School.

Mr. Reel is also a LEED accredited professional for sustainable development and earned the William Pressley Award for research in education for his work in environmental sustainability in schools. In 2004, Mr. Reel was awarded the U.S. Department of Education Presidential Teacher Award. Mr. Reel earned his B.A. in Human Biology and his M.S. in Biology from Stanford University.

About Research & Education Association

Founded in 1959, Research & Education Association (REA) is dedicated to publishing the finest and most effective educational materials—including study guides and test preps—for students in middle school, high school, college, graduate school, and beyond.

Today, REA's wide-ranging catalog is a leading resource for teachers, students, and professionals. Visit *www.rea.com* to see a complete listing of all our titles.

Acknowledgments

We would like to thank Pam Weston, Publisher, for setting the quality standards for production integrity and managing the publication to completion; John Cording, Vice President, Technology, for coordinating the design and development of the REA Study Center; Larry B. Kling, Vice President, Editorial, for his overall direction; Diane Goldschmidt, Managing Editor, for coordinating development of this edition; Kathy Caratozzolo of Caragraphics for typesetting this edition; and Christine Saul, Senior Graphic Designer, for designing our cover.

In addition, we would like to thank Steve Everett for technically reviewing the manuscript; Marianne L'Abbate for copyediting; Ellen Gong for proofreading; and Terry Casey for indexing.

Chapter 1

Welcome to REA's All Access for AP Environmental Science

A new, more effective way to prepare for your AP exam

There are many different ways to prepare for an AP exam. What's best for you depends on how much time you have to study and how comfortable you are with the subject matter. To score your highest, you need a system that can be customized to fit you: your schedule, your learning style, and your current level of knowledge.

This book, and the free online tools that come with it, will help you personalize your AP prep by testing your understanding, pinpointing your weaknesses, and delivering flashcard study materials unique to you.

Let's get started and see how this system works.

How to Use REA's AP All Access

The REA AP All Access system allows you to create a personalized study plan through three simple steps: targeted review of exam content, assessment of your knowledge, and focused study in the topics where you need the most help.

Here's how it works:

Review the Book	Study the topics tested on the AP exam and learn proven strategies that will help you tackle any question you may see on test day.
Test Yourself & Get Feedback	As you review the book, test yourself. Score reports from your free online tests and quizzes give you a fast way to pinpoint what you really know and what you should spend more time studying.
Improve Your Score	Armed with your score reports, you can personalize your study plan. Review the parts of the book where you are weakest, and use the REA Study Center to create your own unique e-flashcards, adding to the 100 free cards included with this book.

Finding Your Strengths and Weaknesses: The REA Study Center

The best way to personalize your study plan and truly focus on the topics where you need the most help is to get frequent feedback on what you know and what you don't. At the online REA Study Center, you can access three types of assessment: topic-level quizzes, mini-tests, and a full-length practice test. Each of these tools provides true-to-format questions and delivers a detailed score report that follows the topics set by the College Board.

Topic-Level Quizzes

Short online quizzes are available throughout the review and are designed to test your immediate grasp of the topics just covered.

Mini-Tests

Two online mini-tests cover what you've studied in each half of the book. These tests are like the actual AP exam, only shorter, and will help you evaluate your overall understanding of the subject.

Full-Length Practice Test

After you've finished reviewing the book, take our full-length exam to practice under test-day conditions. Available both in this book and online, this test gives you the most complete picture of your strengths and weaknesses. We strongly recommend that you take the online version of the exam for the added benefits of timed testing, automatic scoring, and a detailed score report.

Improving Your Score: e-Flashcards

Once you get your score reports from the online quizzes and tests, you'll be able to see exactly which topics you need to review. Use this information to create your own flashcards for the areas where you are weak. And, because you will create these flashcards through the REA Study Center, you'll be able to access them from any computer or smartphone.

Not quite sure what to put on your flashcards? Start with the 100 free cards included when you buy this book.

After the Full-Length Practice Test: *Crash Course*

After finishing this book and taking our full-length practice exam, pick up REA's *Crash Course for AP Environmental Science*. Use your most recent score reports to identify any areas where you still need additional review, and turn to the *Crash Course* for a rapid review presented in a concise outline style.

REA's Suggested 8-Week AP Study Plan

Depending on how much time you have until test day, you can expand or condense our eight-week study plan as you see fit.

To score your highest, use our suggested study plan and customize it to fit your schedule, targeting the areas where you need the most review.

	Review 1–2 hours	Quiz 15 minutes	e-Flashcards Anytime, anywhere	Mini-Test 30 minutes	Full-Length Practice Test 2 hours, 15 minutes
Week 1	Chapters 1–3		Access your e-flashcards from your computer or smartphone whenever you have a few extra minutes to study. Start with the 100 free cards included when you buy this book. Personalize your prep by creating your own cards for topics where you need extra study.		
Week 2	Chapter 4	Quiz 1			
Week 3	Chapter 5	Quiz 2			
Week 4	Chapter 6	Quiz 3		Mini-Test 1 (The Mid-Term)	
Week 5	Chapter 7	Quiz 4			
Week 6	Chapter 8	Quiz 5			
Week 7	Chapters 9–10	Quizzes 6 and 7		Mini-Test 2 (The Final)	
Week 8	Review Strategies in Chapter 2				Full-Length Practice Exam (Just like test day)

Need even more review? Pick up a copy of REA's *Crash Course for AP Environmental Science*, a rapid review presented in a concise outline style. Get more information about the *Crash Course* series by visiting *www.rea.com*.

Test-Day Checklist

✓	Get a good night's sleep. You perform better when you're not tired.
✓	Wake up early.
✓	Dress comfortably. You'll be testing for hours, so wear something casual and layered.
✓	Eat a good breakfast.
✓	Bring these items to the test center: • Several sharpened No. 2 pencils • Admission ticket • Two pieces of ID (one with a recent photo and your signature) • A noiseless wristwatch to help pace yourself
✓	Arrive at the test center early. You will not be allowed in after the test has begun.
✓	Relax and compose your thoughts before the test begins.

Remember: eating, drinking, smoking, cellphones, dictionaries, textbooks, notebooks, briefcases, and packages are all prohibited in the test center.

Chapter 2

Strategies for the Exam

This is your first time taking the AP Environmental Science (APES) exam. Of course it is a mystery to you. However, for those of us who have watched a lot of students take this exam and have reflected on this material for many years, we can see that there are some important patterns from previous exams that will help you in your preparation for the exam. These observed patterns are outlined and analyzed in this chapter. Use this chapter to acquaint yourself with what is on the test and how to approach the multiple-choice and free-response sections. Then, use the quizzes, mini-tests, and full-length practice test that accompany this book to check your readiness for the exam. Your comfort with the style of questions described here will contribute to your success when you take the AP exam.

Multiple-Choice Questions

You will have 90 minutes to answer 100 questions. That means that you will have about 54 seconds—or just under a minute—to read and answer each multiple-choice question. The 100 multiple-choice questions on the APES exam typically take one of the following formats.

- **Most multiple-choice questions give one question and five answer options, such as:**

    Producers in a forest ecosystem produce an amount of biomass and energy that is always less than what is used by consumers in the next trophic level. This is an ecological application of which of the following laws or principles?

    (A) Henry's Law
    (B) First Law of Thermodynamics
    (C) Second Law of Thermodynamics
    (D) The Precautionary Principle
    (E) Rule of 70
    (Answer: C)

- **Some questions provide different possible answers, where more than one right answer is possible:**

    The Coriolis Effect is created by the combined effect of which of the following Earth forces?

    I. The rotation of the Earth
    II. Atmospheric convection
    III. Thermohaline ocean currents

    (A) I only
    (B) II only
    (C) I and II only
    (D) II and III only
    (E) I, II, and III
    (Answer: C)

- **Some questions involve simple calculations, which are possible to answer without using a calculator:**

    A city has a population of 1 million and an annual growth rate of 3.5 percent. If the growth rate remains constant and there is no immigration or emigration, what will be the size of this city's population in 60 years?

    (A) 1.5 million
    (B) 2.0 million
    (C) 3.0 million
    (D) 4.0 million
    (E) 8.0 million
    (Answer: D)

- **Some multiple-choice questions provide the same answers for a number of questions:**

    Questions 1–3 refer to the answers listed below. Select the one lettered choice that best fits each statement.

    (A) Carbon dioxide
    (B) Soluble carbonate ions
    (C) Lignite coal

(D) Limestone

(E) Organic molecules

1. Form of carbon most responsible for global warming (Answer: A)

2. Sedimentary rock formed from calcareous organisms (Answer: D)

3. In equilibrium with carbon dioxide in the oceans (Answer: B)

Strategies for Success on Multiple-Choice Questions

- **Remain calm.** You will remember more and work faster if you are relaxed and confident.

- **Pace yourself.** Have you completed 11 questions after 10 minutes? 22 questions after 20 minutes? Remember, you have 54 seconds to answer each multiple-choice question.

- **Move to the next question if you get stuck.** There is no guarantee that all the easy questions are at the beginning of the test. If you're stuck, go on to the next question. If you have time, go back.

- **If you skip a question, be sure to also skip the answer bubble on the answer sheet.** Periodically, check to be sure you are filling in the bubble that corresponds to the problem number that you intend to answer.

- **Use the process of elimination when you are unsure of an answer.** Right answers tend to be longer and involve more words. Be cautious of answers that use the words "never" or "always"—few things are so absolute and these options can usually be eliminated. It is often difficult for test writers to dream up wrong answers, so answer options that are far-fetched or silly can usually be eliminated immediately.

- **If you don't know the answer, guess.** If you need to guess, you can often eliminate clearly unqualified answer options first, and then guess from the remaining, more reasonable answer options. As odd as it sounds, your first guess tends to be correct most often. If you are guessing on a question, don't change your answer unless you are certain you have remembered the correct answer.

- **For longer questions, underline important terms and the sentence that actually asks a question.** This is a focusing strategy to help you know what is expected of you if you face one of those rare, paragraph-long multiple-choice questions.

Free-Response Questions

As you review the material in the following chapters, keep in mind that questions in the free-response section are often "synthesis" style questions that incorporate multiple topics and require some level of analysis and critical thinking.

In this section, you will see five example questions that are patterned after past APES free-response questions. After you have studied the material in this book, return to this section to practice answering each free-response question below. Allow yourself *20 minutes to answer each question*. Then use the scoring answer sheet after the question to score yourself. If you miss points, use the answer sheet to see which chapter in this review book pertains to that question. Then review the chapter to be sure you understand the material.

Many topics can be covered on the free-response section of the exam. The five question types here represent the very types of questions that have most frequently appeared on the APES tests in recent years. Although other types of questions may appear on the exam, practicing these particular free-response questions will build your confidence and enhance your ability to succeed on the APES exam.

Document-Based Questions

Between 2006 and 2012, we can see that every first question on the free-response section provided an article, interview, or some other type of excerpted publication—often from the fictitious town of Fremont—that gave information for you to read and interpret in order to form your answers. So let's now visit our own fictitious town of Glenmont and see what we can learn about how the APES DBQs work. As you read each question, read actively—underlining information, facts, or opinions.

GLENMONT EXAMINER
To Frack or Not to Frack

After an active public debate yesterday, the City Council for the City of Glenmont will vote today on whether to sell or lease city land to drilling companies who will use the process of hydraulic fracking to extract natural gas. Glenmont sits on an ancient geological formation that contains natural gas embedded within the shale. However, the process of hydraulic fracking pushes previously unavailable natural gas toward wells so that it can be extracted. During the process, millions of gallons of water, sand, and chemicals are pumped into the ground at high pressure to release the natural gas. The water and chemicals either remain in the ground or are removed and disposed of later.

Those who spoke in favor of the process mentioned how important the income would be to the area. Taxes would support schools and the city government, and the mineral rights owners would receive a 12.5% royalty. Those who opposed the vote cited "the precautionary principle," and were worried about air and water pollution. Some suggested that drillers set up a bond for cleanup purposes, similar to that required of surface mining operations in the state.

(a) Discuss two environmental benefits to using natural gas over coal for the generation of electrical power.

(b) Describe "the precautionary principle." How does it relate to this issue?

(c) What reasons were given FOR the use of hydraulic fracking to obtain natural gas?

(d) Who gets the money if the city decides to allow sales and leases for hydraulic fracking? Does the landowner get the money?

(e) Outline three types of pollution that may be associated with fracking.

(f) Which law requires surface mining concerns to post a bond for reclamation once mining operations are complete?

Scoring Sheet

Solutions	Scoring	Chapter to Review
(a) The combustion of natural gas does not emit oxides of sulfur and mercury vapor, both of which would otherwise be produced in the combustion of coal. Natural gas combustion will still produce oxides of nitrogen—which contributes to acid rain, and carbon dioxide—which is a major greenhouse gas.	+1 sulfur +1 mercury	Chapter 8: Energy Resources and Consumption Chapter 9: Pollution
(b) The precautionary principle states that if an action poses a possible risk to public or environmental health, the burden of proof that it is not harmful is on those who wish to take the action. In this case, the burden of proof rests on the drillers to know how much pollution will be created, and that they will clean up any pollution created.	+1 statement of principle +1 correct application to this situation	Chapter 7: Land and Water Use
(c) Reasons FOR fracking were focused on money.	+1 for money	Chapter 7: Land and Water Use
(d) Money is received by the city, schools, and mineral rights holders. Landowners only get royalties if they hold the mineral rights.	+1 for identifying all three recipients. −1 for stating "landowners"	Chapter 7: Land and Water Use
(e) Three types of pollution include: 1. Groundwater 2. Solid waste 3. Air	+1 for each type mentioned	Chapter 9: Pollution
(f) Surface Mining Control and Reclamation Act (SMCRA)	+1 for correct acronym	Chapter 7: Land and Water Use

GLENMONT EXAMINER

Use of Oil in Glenmont Leads to Cancer in Bangladesh

Glenmont Friends of the Earth (GFE), an advocacy organization that monitors environmental issues worldwide, is concerned about the ship breaking and recycling projects in Bangladesh. GFE reported on their website that hundreds of ships at the end of their usefulness make it to these southern Asia shores each year to be dismantled. However, workers die each year from cancer due to exposure to toxic materials in the ships, or die or are maimed as a result of hazardous working conditions. GFE asserts that this recycling economy takes advantage of these workers, who take personal risks because they are desperate for an income.

Most of the ships that come to Bangladesh are oil tankers. Even when the ship is broken apart and the parts sold for scrap, each ship typically leaves behind over a million gallons of oil and several million gallons of sludge. This carcinogenic mixture remains close by and creates the new working environment as workers prepare to accept the next ship. Other toxic compounds that surround workers during the ship-breaking process include asbestos, poly-chlorinated biphenyls, heavy metals, human sewage, acids, and volatile organic compounds.

GFE points out that the traffic of toxic ships to this developing country is a direct result of western thirst for international products and oil, since old oil tankers make up a large proportion of the ships coming to Bangladesh. GFE director, Uli Weiss, states, "Our actions in Glenmont can either contribute to or work against the injustices that are occurring in Bangladesh."

(a) Identify and explain TWO examples of how choices made in Glenmont can "contribute to or work against the injustices" occurring in Bangladesh.

(b) What is a common term that is used for the exploitation identified in this article? Give another real example of analogous environmental exploitation that has occurred in history, or continues to occur today.

(c) What costs are not being included in the "true cost analysis" of owning and operating the ships mentioned in this article? Devise a mechanism of incorporating these external costs within the true cost of owning the ships.

(d) Explain the concept of "dose" with respect to toxic substances. Include in your explanation the relationship between the likelihood of cancer and the amount of exposure to a carcinogen. How are these concepts related to the information presented in this article?

(e) Of the list of toxic chemicals listed in this article, pick TWO and describe the type of toxicity it represents to humans.

Scoring Sheet

	Solution	Scoring	Chapter to Review
(a)	Students may list two actions that involve buying local merchandise and/or using fewer oil-based products.	+1 for each of three items mentioned	Chapter 8: Energy Resources and Consumption
(b)	Either "environmental racism" or "toxic colonialism" may be used. One of any number of examples may be used. A common example might be the dumping of incinerator ash on foreign shores.	+1 for term +1 for example	Chapter 9: Pollution
(c)	The health costs of the workers and the cost of cleaning up the oil sludge on the shores are not being considered by the shipowners. This cost might be covered either through legal action or by having the shipowners pay a fee before the ship is accepted by the port.	+1 health and/or environmental costs +1 for a reasonable strategy to internalize the cost	Chapter 7: Land and Water Use
(d)	Dose is defined as the amount and length of time of exposure. In this case, the fact that the carcinogens remain on site increases the length of time workers are exposed, thus increasing their chances of getting cancer.	+1 Definition of dose +1 Application to this article	Chapter 9: Pollution
(e)	asbestos — carcinogen PCBs — carcinogen Heavy metals — neurotoxin, carcinogen Human sewage — pathogen Acids — corrosive VOCs — carcinogen	+1 for each correct +2 maximum	Chapter 9: Pollution

Energy Questions

Between 2007 and 2012, the free-response sections on the AP Environmental Science exam have notably contained a lengthy problem that pertains to energy, and involves some level of solving problems with units.

Glenmont's coal-fired electrical power plant uses coal and produces power at the rate shown in the chart below.

Average daily use of coal	4,500 tons coal/day
Energy from each pound of coal	5,000 BTU/lb coal
Number of BTUs in 1.0 kWh	3,400 BTU/kWh
Efficiency of each energy transformation	80%
Amount NO_x prevented by 2-staged burners	0.75 lb NO_x/million BTUs

(a) How many kWh of electrical energy does the plant produce each day?

(b) Beginning with coal, identify the energy transformations that take place in order to produce electricity in the power plant. Calculate the total efficiency of the power plant.

(c) What are the four major types of air pollution from a power plant, and what impact do each of these types of pollution have on ecosystems around the plant and globally?

(d) Describe how two-staged burners prevent the formation of NO_x emitted by the plant.

(e) How many pounds of NO_x are prevented per day by the operation of two-staged burners?

Scoring Sheet

Solution	Scoring	Chapter to Review
(a) 1.3×10^7 kWh = 4500 tons $\times \dfrac{2000 \text{ lb}}{\text{ton}} \times \dfrac{5000 \text{ BTU}}{\text{lb}} \times \dfrac{1 \text{ kWh}}{34000 \text{ BTU}}$	+1 for correct units +1 for correct answer	Chapter 7: Land and Water Use
(b) Steps could include: • coal • combustion heats water • water pushes turbine • turbine turns generator • turbine produces electricity • Efficiency: $0.8^4 = 0.24$, or 24%	+1 for each step +2 maximum for steps +1 for correct calculation +3 maximum for question	Chapter 7: Land and Water Use
(c) NO_x — acid rain SO_x — acid rain Mercury — neurotoxin Carbon dioxide — greenhouse gas, global warming	+1 for each +4 maximum	Chapter 7: Land and Water Use
(d) At the first stage, combustion takes place at high temperatures with little oxygen. At the second stage, combustion takes place in an oxygen-rich, low-temp environment. Results in lower NO_x production.	+1 for correct answer	Chapter 7: Land and Water Use
(e) 1.3×10^7 kWh = $\dfrac{0.75 \text{ lb}}{10^6 \text{ BTU}} \times \dfrac{3400 \text{ BTU}}{\text{kWh}} \times 1.3 \times 10^7$ kWh	+1 for correct answer	Chapter 7: Land and Water Use

Climate Change Questions

Every test in the last six years has had at least one question on the relationship between carbon and climate change. Many of these questions involve a chart, graph, or map. Most often, these questions draw some relationship between temperature, carbon dioxide in the atmosphere or dissolved in the ocean, and carbon sequestered in some other form on the planet.

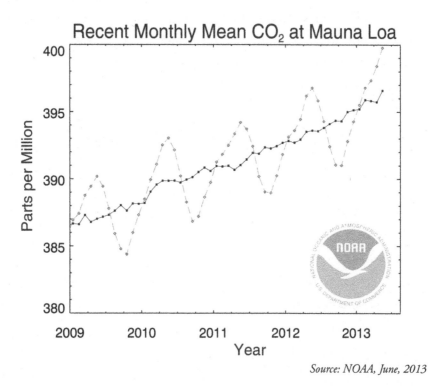

Source: *NOAA, June, 2013*

(a) Why does the annual cycle of atmospheric carbon dioxide oscillate?

(b) What is the increase in average atmospheric values of carbon dioxide between 2008 and 2012? Express your answer in parts-per-billion (ppb).

(c) Identify how the following Earth systems change in response to increased levels of atmospheric carbon dioxide.

    i. Global temperature

    ii. Carbon dioxide dissolved in the ocean

    iii. Sea pH level

    iv. Sea level height

(d) As humans continue to need wood from forests for building structures and building fires, global forests are decreasing. What is the relationship between deforestation and atmospheric carbon dioxide?

(e) A business constructed a 0.5 MW solar array in an area that receives about 200 days of sunshine per year, with an average of 10 hours of sunshine on a sunny day.

   i. What is the total number of MWh produced by this solar array in one year?

   ii. As a result of using this solar array for 10 years, how many kg of carbon dioxide has been prevented from going into the atmosphere? (Assume that burning coal to produce 1.0 MWh of energy will release about 2000 pounds of carbon dioxide into the atmosphere.)

Scoring Sheet

	Solution	Scoring	Chapter to Review
(a)	There is less photosynthesis in the winter months.	+1	Chapter 10: Global Changes
(b)	About 8 – 10 ppm, or 8000 – 10000 ppb	+1	Chapter 9: Pollution
(c)	i. Global temperature increases ii. Carbon dioxide increases iii. Sea pH level decreases iv. Sea level height increases	+1 each +4 maximum	Chapter 10: Global Changes
(d)	Forests sequester carbon by using up carbon dioxide in photosynthesis. As the forests disappear, levels of atmospheric carbon dioxide increase at an accelerated rate.		Chapter 10: Global Changes
(e)	i. $3400 \text{ MW} \times 200 \text{ days} \times \frac{10 \text{ h}}{\text{day}} = 1000 \text{ MWh}$ ii. $1000 \text{ MWh} \times \frac{2000 \text{ lb}}{\text{MWh}} = 20,000,000 \text{ lb}$	+0.5 correct units +1 for correct answer	Chapter 8: Energy Resources and Consumption

Map-based Questions

OHIO RIVER BASIN
Location of the 18 Power Plants

The Ohio River composes a vast network of coal-fired power plants that is supported by a network of dams and locks on the Ohio River. The dams and locks, in combination with steady dredging of a channel in the river, allows coal barges to transport coal from Kentucky and Tennessee along the Ohio River, to the power plants. Additionally, the Ohio River serves as a heat sink, as the river water is diverted into the condensers of the power plant to cool the steam so it can be re-used in the generation process.

(a) How does the channelization of the Ohio River for the sake of increasing ship and barge traffic affect flooding during high run-off years?

(b) To reduce the impact of flooding, some towns along the Ohio River have constructed dikes. What affect does that have downstream?

(c) Identify one type of water pollution and three types of air pollution that are created by these power plants.

(d) For each of the types of pollution you mentioned in (c), identify one method to treat or remove that pollutant.

(e) Consider the location of the Ohio River network of power plants in relation to the circumpolar vortex that sweeps across the United States.

  i. Describe the movement of air across the U.S. as a result of the circumpolar vortex.

  ii. From this information, and from what you know about pollution from coal-fired plants, make a hypothesis about pollution conditions in specific locations in the U.S. not shown on this map that would be affected by these power plants.

  iii. Design an experiment by which you could test this hypothesis. Include a "control," and identify what variables you would measure.

Scoring Sheet

Solution	Scoring	Chapter to Review
(a) Flooding increases	+1	Chapter 7: Land and Water Uses
(b) Dikes increase channelization, which increases downstream flooding.	+1	Chapter 7: Land and Water Uses
(c) Water: thermal pollution Air: NO_x, SO_x, mercury, CO_2	+1	Chapter 9: Pollution
(d) Thermal — cooling ponds NO_x — two-staged burner SO_x — fluidized bed combustion, flue gas desulfurization CO_2 — carbon sequestration, as in planting forests	+1	Chapter 9: Pollution
(e) i. Circumpolar vortex moves air from west to east across the U.S.	+1	Chapter 4: Earth Systems
ii. More acid rain, mercury, and carbon dioxide would end up on the eastern seaboard of the U.S.	+1	Chapter 4: Earth Systems
iii. Measure acid rain, mercury, and carbon dioxide levels at various times in various eastern cities. Correlate these readings with weather data to see if the readings increase when a particular air mass moves directly from the Ohio River basin to the study site. The control is the lowered level at the study site when an air mass does not bring air directly to that location.	+1	Chapter 4: Earth Systems

Strategies for Success on Free-Response Questions

- **Pace yourself.** You have 90 minutes to answer the four free-response questions. Allow yourself about 20 minutes for each of the four questions. That leaves you ten minutes to re-read your answers. If your answers involve multiplication, you can check your arithmetic.

- **Read through the entire question before starting to write down an answer.**

- **Underline the sentence that actually asks a question.** Free-response questions are longer and may include information that distracts you from the question. You can focus your thoughts if you underline what is actually required of you.

- **Focus your answer on answering the question that is asked.** Take it one question part at a time. Don't worry how long the question might be. This will focus your thinking and keep you from feeling intimidated by longer questions or large data tables.

- **Be neat.** Remember that the person reading your answer is a human being, reading thousands of the same answer. Make it easy for him/her to see your answer by being neat. Underline answers (numbers AND units) in problems so that the reader can find your answer easily.

- **On problems, show your work. You might multiply incorrectly, but do the problem correctly.** Showing your work maximizes your chance of earning partial credit.

- **Write clearly.** Use clear, full sentences. If there is a key word within a sentence, don't hesitate to underline the word to make sure that the reader knows you have included it. Make sure that you have connected the key words of your answer to the actual question that is asked. Don't require the reader to make any assumptions, or logical leaps—make them yourself, within your writing.

- **Be prepared to draw and label any graphs that you have studied.** They may provide good examples on free-response questions. It is okay to have graphs, chemical reactions, or other types of communication within your answer, in addition to the written sentences.

- **Both knowledge of facts and careful reading of the documents yield the best answer.** There is no substitute for your knowledge of the topics on the exam. In addition, be sure to read the documents found in the exam carefully and apply your knowledge. Reading the document might offer you clues about how to respond to the questions.

Strategies for Overall Preparation

- **Use this book to summarize topics and practice answering questions.** Repeat a review of topics represented in practice problems that you miss. Most courses and textbooks include much more in some topic areas than you need for the AP exam, and neglect to cover enough in other areas. This review book will help you focus on what is important for the AP exam.

- **Keep connecting topics from one chapter to topics in other chapters.** Environmental science is unique in that the topics are tightly interconnected. Part of your task is to learn and remember each topic individually, and the other part of your task is to recognize and understand how they are connected.

- **Study with another student or in a group.** Talking about the topics, quizzing each other, and holding each other accountable can make the material seem easier and help all of you remember key points.

- **Plan several days to prepare for the test.** Cramming does not work. Being confident and knowledgeable about the material is your best ally for passing the exam.

- **Acquaint yourself with past AP test questions and practice questions.** Use this book and the online practice components for thorough preparation. In addition, take the time to look at the APES material on the College Board website.

- **Plan your schedule well enough to get good sleep during the week before the test.** You'd be surprised by how much this can help!

Basic Concepts and Skills Review

Chapter 3

Students who take an AP Environmental Science (APES) course come from a wide range of backgrounds and may or may not have taken previous science courses. This chapter reviews material and skills that you are expected to know for the AP Environmental Science exam but that you won't see on the College Board's AP Environmental Science syllabus. Areas that are covered here and that are typical expectations on the AP Environmental Science exam include basic principles of matter, the basics of multiplying and dividing numbers with units of measurement, exponents, and percentages.

Review of Principles of Matter

The particles in the environment are composed of matter. Questions on the AP exam may contain assumptions about some aspect of matter that may not officially be a part of the AP Environmental Science syllabus but that you are expected to know. The following are examples of the level of understanding of matter, and the AP Environmental Science topics where these concepts tend to show up.

Atoms are the smallest unit of elements that are listed on the periodic table. The nucleus of an atom is composed of neutrons and protons, while electrons exist outside the nucleus. It is important to remember the atomic nature of matter in order to understand several important APES topics, such as pollution, energy, and the biogeochemical cycles.

Isotopes are atoms that have the same number of protons but a different number of neutrons. This is important when discussing radiation and radioactive decay in an AP Environmental Science course. For example, iodine is an element that humans need to live. However, if iodine is bombarded with extra neutrons, such as the case when there is a nuclear reactor accident, those extra neutrons make an isotope of regular iodine that is radioactive and harmful to human health.

Ions are atoms or groups of atoms that carry an electric charge. Negative ions have more electrons than protons. Positive ions have more protons than electrons. This is important in environmental science because the biogeochemical cycles discussed later often involve matter that is organized as ions.

TEST TIP

Be sure you are able to identify ions and compounds that are key pollutants, participants in pollution control, and members of one of the biogeochemical cycles. These ions and compounds are listed throughout this book.

Compounds are collections of atoms held together by chemical bonds. Ionic compounds, such as salts, use ionic bonds to hold metal atoms together with nonmetal atoms. Molecules use covalent bonds to hold nonmetal atoms together, such as with organic molecules.

Acids are molecules that are able to donate protons when put in water. pH is a measure of acidity. The pH value is the negative exponent of the concentration of hydrogen ions. A pH below 7.0 is acidic; a pH above 7.0 is basic.

Chemical reactions show how a group of reactants change into products. While you need not remember all the reactions from biological processes, such as photosynthesis, you should be able to recognize the context of a few key chemical reactions. These key reactions are identified in this book.

Conservation of matter states that matter is neither created nor destroyed when it is rearranged in a chemical reaction. A chemical reaction shows one or more reactants turning into one or more products. At the atomic level, the conservation of matter is expressed in a chemical reaction.

TEST TIP

Conservation of matter is an important concept in several environmental topics. It is the basis of the nutrient cycles, where all mass is accounted for as it moves through the cycle. With respect to pollution, conservation of matter is important because toxic materials are matter that is conserved—that is, it never "goes away." For example, a toxin will pass through the food web from prey to predator and still be toxic.

States of matter refer to any one of three states in which matter exists commonly on Earth: solid, liquid, or gas. Matter needs to absorb energy and increase the thermal motion of its atoms in order to pass from solid to liquid, or from liquid to gas. Several AP Environmental Science topics refer to different states of matter, such as climate and weather, and the different forms pollution can take.

Review of Math Skills

The AP Environmental Science exam does require that you do some calculations—and without a calculator no less! Don't be intimidated by these problems—a little studying goes a long way. The fact that no one can use a calculator means that any calculations must be very simple. Most AP Environmental Science textbooks do not spend enough time showing students how to answer AP Environmental Science exam-style calculations, so reviewing this section—as well as the chapters on "Populations" and "Energy Resources and Consumption"—are very important preparation for taking the exam.

> **TEST TIP**
>
> If you're taking a long time to do arithmetic on an exam problem, you're probably not doing it correctly. Take the time to rethink your approach to the problem.

Exponential notation helps to express numbers that are either very large or very small, as is often the case in the natural world. It is also an easy way to multiply and divide numbers quickly.

1. Numbers expressed using exponential notation are composed of a group of significant digits multiplied by a power of ten. For example,

   420 can be expressed as 4.2×10^2

   6,890,000 can be expressed as 6.89×10^6

   0.000045 can be expressed as 4.5×10^{-5}

2. When adding or subtracting numbers with exponents, convert all the numbers to the same power of ten, and then add or subtract just the significant digits. For example,

   $400,000 + 2.2 \times 10^6 =$

   $(0.4 \times 10^6) + (2.2 \times 10^6) = 2.6 \times 10^6$

3. When multiplying numbers with exponents, simply add the powers of ten. For example,

$$(4.5 \times 10^3) \cdot (2.0 \times 10^5) = 9.0 \times 10^8$$

$$(2.5 \times 10^6) \cdot (3.0 \times 10^7) = 7.5 \times 10^{13}$$

$$(1.2 \times 10^{-2}) \cdot (4.0 \times 10^{-3}) = 4.8 \times 10^{-5}$$

4. When dividing numbers with exponents, simply subtract the powers of ten. For example,

$$(6.0 \times 10^5) / (2.0 \times 10^2) = 3.0 \times 10^3$$

$$(9.0 \times 10^8) / (3.0 \times 10^5) = 3.0 \times 10^3$$

$$(8.0 \times 10^{-2}) / (4.0 \times 10^6) = 2.0 \times 10^{-8}$$

TEST TIP

Much of the multiplication and division on the AP Environmental Science exam can be done using exponents. Remember, when multiplying, add the exponents; when dividing, subtract the exponents.

Percentages are frequently found on the APES exam.

1. Taking percentages is found by creating a fraction and multiplying by 100, which takes that fraction and puts it in terms of "per hundred." Percentages are used particularly in problems having to do with population and energy.

EXAMPLE What is the total percent efficiency of an engine that burns 30 joules of chemical energy to produce 20 joules of mechanical energy?

SOLUTION $\dfrac{\text{output}}{\text{input}} \times 100\% = \dfrac{20 \text{ joules}}{30 \text{ joules}} \times 100\% = 66\%$

2. Adding and subtracting percentages shows up primarily in population problems, where a percent growth is found by subtracting a death rate from a birth rate.

> **EXAMPLE** A population has a birth rate of 7 new births per hundred people in the population. It has a death rate of 5 deaths per hundred people. Assuming no immigration and emigration, what is the total growth rate in the population?

> **SOLUTION** 7% − 5% = 2%. The population has a 2% growth rate.

Problem solving with units is commonly used in at least one place in the typical APES exam, usually in the free-response section. Tracking units in problem solving and making sure that the units of one side of an equation are the same as the units on the other side is called *dimensional analysis*.

Don't be intimidated by solving problems! Just take it one step at a time. You've probably solved these problems in many ways in the past, but with more familiar units. (For example, *miles per hour* divides the number of *hours* into the number of *miles* to get a number and its corresponding units of velocity.) If you took a chemistry class, this method is probably very easy for you. Even if you didn't take a chemistry course, multiplying and dividing with units simply requires some practice, paying attention to the problem, and some common sense.

These problems typically show up in energy problems, particularly problems that convert one type of energy to another. Here is a series of steps that you can use to answer these questions without relying entirely on your common sense:

STEP 1. Read the problem carefully. Underline or write down any number and the corresponding unit.

> How many joules of heat are produced from 2,000 lbs. of coal? There are 1054 joules in one BTU, and 1.0 lb. of coal produces 5,000 BTUs.

STEP 2. Write down the units called for in the answer with an "=" next to it.

> x joules =

STEP 3. Write down the given information on the right side of the "=" sign from the answer.

$$x \text{ joules} = 2{,}000 \text{ lbs. of coal}$$

STEP 4. Convert the given information into the units of the answer, remembering that any unit or number that exists in both the numerator (on "the top") and the denominator (on "the bottom") will cancel each other out.

$$x \text{ joules} = \frac{5000 \text{ BTUs}}{1 \text{ lb coal}} \times \frac{1054 \text{ joules}}{1 \text{ BTU}}$$

STEP 5. Multiply all the numerators together, and divide by all the denominators. The remaining units on both sides of the "=" sign should be the same.

$$1.1 \times 10^{10} \text{ joules} = 2000 \;\cancel{\text{lbs of coal}}\; \times \frac{5000 \;\cancel{\text{BTUs}}}{1.0 \;\cancel{\text{lb coal}}} \times \frac{1054 \text{ joules}}{1 \;\cancel{\text{BTU}}}$$

> **TEST TIP**
>
> Typically, the AP Environmental Science exam will not require using more than one conversion factor at a time to solve a problem. However, some free-response questions will ask a series of questions, where the answer for the first question becomes the given information of the next question. Using dimensional analysis is a way to feel confident about the most difficult questions that can be asked on the exam.

Analyzing Scientific Data

Maps

Many AP Environmental Science topics refer to a particular physical or geopolitical situation that occurs on the Earth. Such topics may range from who produces the most coal, to where low-pressure zones exist and cause deserts, to the impacts of El Niño, to where different types of crustal plate boundaries exist. Students will need to be able to identify where on a global map such events occur. Some knowledge of global human and physical geography is essential. Below is an example question that could use a world map.

Questions 12–15 refer to the following map. Select the one lettered choice that best fits each statement.

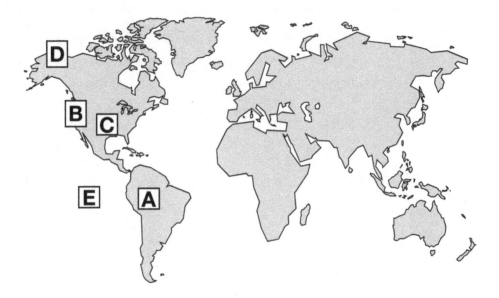

12. Area of the world that contains Taiga forests

13. Ogallala aquifer

14. Rain shadow exists on the eastern side of these mountains.

15. Tropical rainforests exist here.

The correct answers are: 12. (D), 13. (C), 14. (B), and 15. (A).

Graphs

There will likely be several graphs on the AP Environmental Science exam that may show up in both the multiple-choice and free-response sections of the test. While some questions may not require you to use information from the graph, in other cases you will be asked to use information from the graph to answer a specific question. As you interpret the graph, here are some questions to guide you.

1. Notice the units of the question being asked of you, and the units of the vertical and horizontal axes of the graph.

2. Remember that the dependent variable is the vertical, or *y*-axis; and the independent variable is the horizontal, or *x*-axis.

3. Notice whether the graph is a bar graph or a line graph.

   a. Line graphs show a continuous relationship between the data points. In the chapter called "Pollution," you will review a dose-response relationship expressed on a line graph, such as the following.

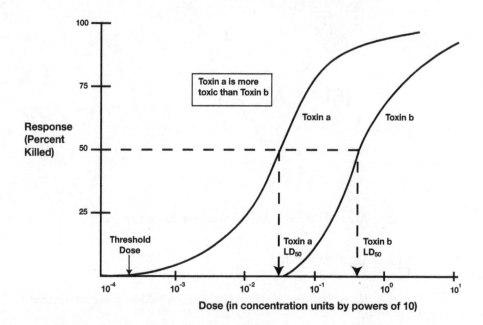

Figure 3.1 Typical sigmoidal dose-response curve. The horizontal axis is usually a logarithmic scale. The vertical axis represents the response—which could refer to anything from being healed from a disease to contracting cancer. When the response recorded on the vertical axis is death, the dose that kills 50% of the subjects is called the lethal dose 50%, or LD_{50}.

   b. Bar graphs show a discrete relationship between the data points. On the next page is an example of a bar graph showing the water quality of rivers, lakes, and salt marshes.

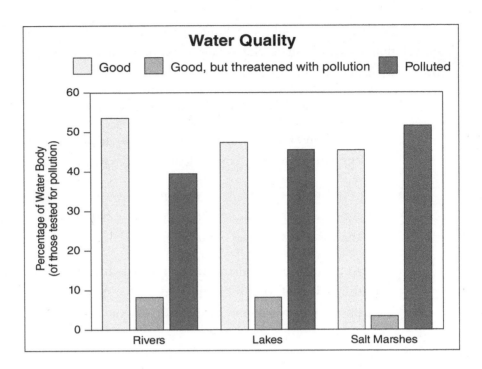

Data Tables

The AP Environmental Science exam frequently uses tables to provide information, and you will need to use information from the table to answer a question. These tables could occur on either the multiple-choice or free-response sections of the exam. Below is an example of the type of data table and questions that you might see on the exam.

Country	Crude Birth Rate	Crude Death Rate	Infant Mortality Rate	Per Capita Income (US dollars)
United States	14	8	6.7	42,000
China	12	7	27	6,500
Kenya	43	19	100	1,000

1. Use data from the table above to contrast two countries at different stages of a demographic transition.

2. Assuming no immigration and emigration, compare the percent population growth in Kenya and the United States.

3. Describe two environmental impacts from overpopulation.

Pause for a minute and make some observations about this type of question. These observations can help you develop stronger test-taking habits and be more aware of opportunities to deliver a strong answer.

1. Not every question requires the data table to answer the question.

2. The question may require several skills. This question requires calculation, graph interpretation, memory, and articulating "big picture" concepts.

3. You need to be familiar with the units involved with common environmental definitions. For example, the units for *crude birth rate* are the number of births *per thousand people* in the population.

4. There is much more information provided than what you need to answer the questions. Part of what is being tested is your ability to discern what information is valuable in constructing a clear, succinct answer.

Review of Experimental Design

Increasingly, the College Board expects that students have not only performed experiments in the field and lab, but have also had the chance to design an experiment. It is not as difficult as it may seem; we all use experimental design to figure things out. We ask ourselves a question, develop a hypothesis, and then apply a series of tests to get the answer to that question (and see how our assumption plays out). From our observations, we draw a conclusion. This is simply the scientific process.

The AP exam often includes a question that asks the student to design an experiment. These questions often show up in toxicology or ecology questions.

Hypothesis represents the question the experimenter is trying to answer, or a statement whose truthfulness will be tested in the experiment. What happens to seed germination when the temperature is changed? The ability of seeds to germinate is measured or observed as the experimenter changes the temperature. Germination becomes the dependent variable, and temperature becomes the independent variable.

Data is the information that is recorded as a result of making an observation or measurement. In the above example, the ability of seeds to germinate will be recorded along with the temperature they experienced. An easy way to record data is on a table that shows how the dependent variable changes as you change the independent variable.

Controls are represented by the variables that are kept constant during an experiment. For example, when growing seeds, we would want the amount of sunlight, water, atmospheric humidity to all be the same with all the plants, so that we would be able to conclude that any differences in germination were due *only* to changes of temperature. The controls help eliminate the possibility that other variables are responsible for a change in the dependent variable.

Data analysis can occur in several ways. If a graph is used, the independent variable should be along the horizontal, or *x*-, axis. The dependent variable should be represented on the vertical, or *y*-, axis. Label the units on each axis. Use a line graph for data that implies a continuous relationship. Use a bar graph to represent data that uses fixed points, and there are no possible points implied between the measured points. In this example, it is implied that seeds could germinate at temperatures between those measured, so a line graph would be the right choice.

Conclusion is the outcome of the experiment, the answer to the question you asked as your hypothesis. In this example, you will find that there is a bell curve of the number of seeds that germinate at different temperatures. (By the way, this bell curve is called a "tolerance curve," which is further described in Chapter 5.)

Chapter 4

Earth Systems

Now that you have developed a sense of the AP Environmental Science exam and reviewed a few basic skills, it's time to jump into the content. Let's begin with reviewing the systems of the Earth that form the basis of the rest of your AP Environmental Science course.

Earth systems are the mechanisms by which solids, liquids, and gases move about the Earth. Also included is the composition of soils. This is a good place to begin your review because all other chapters in some way relate to this chapter: The Earth and its systems are the stage upon which environmental principles occur.

The College Board syllabus indicates that the following topics from this chapter may show up on the AP Environmental Science exam. These topics compose about 10–15% of the total test.

A. **Earth Science Concepts**
(Geologic time scale; plate tectonics, earthquakes, volcanism; seasons; solar intensity and latitude)

B. **The Atmosphere**
(Composition; structure; weather and climate; atmospheric circulation and the Coriolis Effect; atmosphere–ocean interactions; ENSO)

C. **Global Water Resources and Use**
(Freshwater/saltwater; ocean circulation; agricultural, industrial, and domestic use; surface and groundwater issues; global problems; conservation)

D. **Soil and Soil Dynamics**
(Rock cycle; formation; composition; physical and chemical properties; main soil types; erosion and other soil problems; soil conservation)

Global Movement of Matter and Energy

The well-being of all of Earth's ecosystems depend on how matter and energy are dispersed around the planet. The movement of matter around the globe is driven by energy in one of three different ways: radiation, conduction, and convection.

Radiation is energy that is carried by a photon, or light, through space, from one place to another. For example, the sun produces energy in nuclear fusion reactions. That energy is carried through space to the Earth. This is an important principle in AP Environmental Science as we look at global warming and ozone depletion.

Conduction of energy occurs when one particle collides with another and, as a result, energy is transferred. The collision of tectonic plates at plate boundaries is a large-scale example of energy conduction.

Convection occurs when energy-containing particles move from one place to another because of differences in density. The density of the material decreases when warmed. When a material's density changes, it moves from one place to another. Convection currents drive our weather, ocean currents, and the movement of molten lava in the Earth's crust.

The Earth

Earth's layers consist of the **core**, the **mantle**, and the **crust**.

1. The core is composed mostly of nickel and iron.

2. The mantle contains two sublayers, the lower and the upper mantle. The upper mantle is composed of the asthenosphere. The upper mantle and the crust are called the lithosphere. Energy from the Earth's core heats the mantle material, which then rises through convection and eventually comes into contact with the cooler crust, or lithospheric plates. After imparting its energy through conduction, the cooled mantle descends to be reheated by the core. In some places, the rising heated magma punches through the lithospheric plate, creating volcanic activity.

3. The crust is composed of tectonic plates, which slowly move relative to one another.

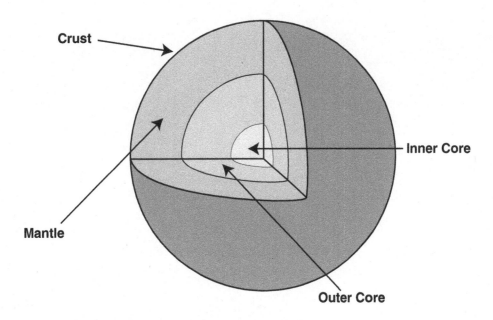

Figure 4-1. Earth's Layers

Plate tectonics is a theory that describes how the Earth's crust is composed of plates that move relative to one another. In 1912, Alfred Wegener proposed the theory of continental drift, which hypothesized that all the continents were originally a single continent that began to break apart about 180 million years ago. It wasn't until the 1960s that scientists put together strong supporting evidence of Wegener's theory, when they found that the mid-Atlantic rift was actually a place where molten lithospheric material was being pushed to the surface of the crust. This upwelling of molten material pushed apart the two plates on either side of the rift. The movement of these plates causes many geological changes, such as creating new types of rocks in the rock cycle, building mountains, and creating volcanoes. The three ways that plates move relative to one another are described by the three types of plate boundaries: divergent, convergent, and transform.

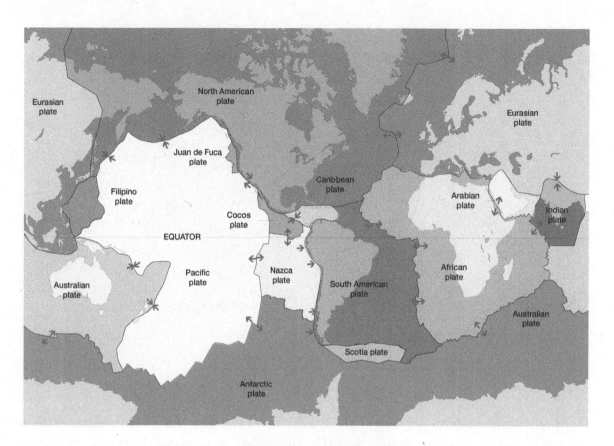

Figure 4-2. Earth's Plates

1. **Divergent boundaries** (also called constructive boundaries or sea floor spreading if the boundary is at the bottom of the sea) occur where magma is pushed up from the mantle and new crust is produced. This upward movement pushes the two plates apart. The Mid-Atlantic ridge—between the African plate and the South American plate—is an example of a divergent boundary.

2. **Convergent boundaries** occur when one plate collides with another plate. These are also called destructive boundaries. The geological result of the collision depends on whether the plate is an oceanic plate or a continental plate. This is significant because continental plates are less dense than oceanic plates.

    An oceanic-continental convergent boundary causes the oceanic plate to dive underneath the continental plate, which causes a deep trench, or "subduction zone." The friction of one plate moving underneath another turns rock into magma, which then rises to the surface to form a volcano on the continental plate. The Nazca plate subducting under the South American plate forms the Andes Mountains.

The convergence of two oceanic plates causes one plate to dive underneath another, which causes a deep trench. The Marianas Trench occurs where the Pacific plate is subducted underneath the Philippine plate.

The convergence of two continental plates forms mountains. Neither is subducted under the other because both plates are less dense and resist being pushed into the asthenosphere. As a result, both plates buckle and push each other upward. The highest mountain range in the world, the Himalayas, was created by the collision of the Indian and Eurasian continental plates.

3. **Transform boundaries** occur when two plates slide past one another, either in the same or opposite directions. This creates an area that is prone to earthquakes, such as the San Andreas Fault in California, which is created as the Pacific plate moves alongside the North American plate.

TEST TIP

For your AP Environmental Science exam, there is no reason to memorize all the major plate names on the Earth's surface. However, it would be helpful to remember the examples of each type of boundary and their location on a map.

The **Geologic Time Scale** for major Earth changes is very long. Scientists currently approximate the Earth's age at more than 4 billion years. With this length of time to allow for change, the movement of continents appears to have been significant, given the current rate of plate movement. For example, the rate at which the Arctic plate moves is about 2.5 centimeters per year. The faster Pacific plate moves at about 15 centimeters per year.

Earthquakes occur when a large amount of earthen material quickly adjusts to a new position. In this process, a large amount of kinetic energy is released and travels through the Earth in waves. The spot where the adjustment takes place is called the "focus." The spot on the surface of the Earth directly above the focus is called the "epicenter."

The **Richter scale** is a logarithmic scale that measures earthquake intensity. For every increase in one unit on the Richter scale, there is a tenfold increase in energy of the earthquake. Smaller earthquakes that are barely felt have a value of about 3.0 on the Richter scale. An earthquake that measures 5.0 on the Richter scale will make it difficult to stand up if one is close to the epicenter. In addition, items will fall off a shelf and landslides may occur under certain surface conditions. A 7.0 earthquake will shake houses off their foundations, create large cracks in the ground, and perhaps cause liquefaction of some soils.

Volcanism is the movement of molten mantle material, called magma, toward the Earth's surface. When the magma cools, it forms volcanic, or igneous rock, described in the section called Soil Resources. Volcanism occurs at mid-plate hotspots, where a plume of molten rock penetrates the crust, and at plate boundaries. The Hawaiian Islands were formed as the Pacific plate moved over a mid-plate hotspot. An example of volcanism at a plate boundary is seen at the edge of the Pacific plate, which extends from Alaska, down the west coasts of North and South America, and over to the eastern coast of Asia.

> **DID YOU KNOW?**
>
> In 1883, Krakatau Island, located on the western edge of the Pacific Ocean's "Ring of Fire," underwent a cataclysmic volcanic explosion. The blast blew a five-mile-wide caldera, or volcanic crater, and created a 140-foot-high tsunami wave, which killed about 34,000 people on low-lying Pacific islands. Earthquakes were felt 500 miles away, and the particulate matter partially blocked the sun and cooled the Earth for more than a year afterward. A similar explosion in 1815 on the island of Tambora decreased average global temperatures by 3°C, which caused famines in North America and Europe.

The Atmosphere

The origin of the atmosphere probably took place in the first few hundred million years of the Earth's more than 4-billion-year history. Originally, the atmosphere was composed of hydrogen and helium, but these gases could not be retained by the Earth's gravitational pull. Volcanic activity produced water vapor, carbon dioxide, sulfur dioxide, chlorine, nitrogen, ammonia, and methane.

Oxygen in the atmosphere was created later in one of two ways:

1. Ultraviolet rays broke up water molecules to form ozone, which decomposed into the oxygen we breathe.

2. The majority of the oxygen in our atmosphere was produced by photosynthetic cyanobacteria between 2.5 billion and 400 million years ago.

3. Oxygen production was later aided by plants.

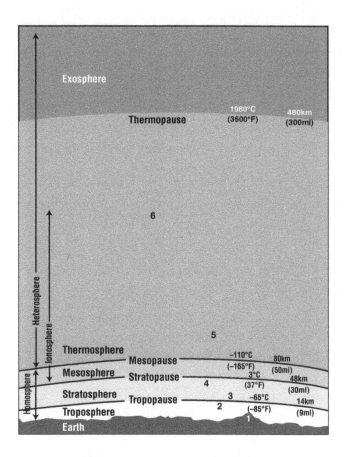

Figure 4-3. Layers of the Atmosphere

Atmospheric layers are defined by differences in the content of air, temperature, and density. The layers closest to the Earth's surface are most dense.

1. The **troposphere** is the layer closest to the Earth and extends upward about 7 miles from the surface. Most of the weather and 75% of the air molecules exist in the troposphere. Generally, temperature decreases with higher altitude in the troposphere.

2. The **stratosphere** is the layer where temperature increases with higher altitude because of the heat emitted when ozone absorbs high-energy ultraviolet (UV) rays radiated from the sun. The ozone layer protects living organisms from damaging radiation. Most jet travel occurs in this layer. The only weather that occurs in the stratosphere is the top portions of very large and energetic thunderstorms. The stratosphere extends from the top of the troposphere to about 30 miles above the Earth's surface.

3. The **mesosphere** exists from the top of the stratosphere to about 50 miles above the Earth. Temperature again decreases with altitude because of the lack of absorption of UV light.

4. The **thermosphere** exists from the top of the mesosphere to about 300 miles above the Earth. Within the thermosphere is the ionosphere, which absorbs high-energy radiation from the sun, causing this layer to increase in temperature with altitude.

5. The **exosphere** is the transition into interstellar space. Most satellites orbit in this layer, and the few gas molecules that exist are typically low-density gases, such as hydrogen and helium, and these only exist at very low pressures.

Composition of the Atmosphere

Table 4-1. Composition of the Atmosphere

Gas	Percent in Atmosphere
Nitrogen	78%
Oxygen	21%
Water vapor	0–4%
Argon	0.9%
Carbon dioxide	0.04%
Neon	0.002%
Helium	0.0005%
Methane	0.0002%
Hydrogen	0.00005%
Ozone	0.000004%

The **flow of energy from the sun to the Earth** is depicted in the following figure. As the radiant energy enters the atmosphere, high-energy UV light tends to be absorbed by the ozone in the stratosphere. Visible light penetrates further. Low-energy infrared light tends to be absorbed by carbon dioxide and water vapor in the troposphere. When the radiation hits the Earth, most is reflected, particularly from surfaces such as water or snow. Darker surfaces, such as forests, absorb more energy. Eventually, all energy absorbed by the Earth is radiated back into space as low-quality heat, some of which is held in by clouds or particulate matter, which helps to rewarm the Earth. (We will discuss this later when we cover the greenhouse effect in Chapter 10.)

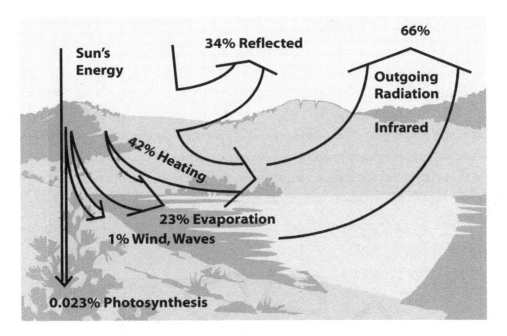

Figure 4-4. Energy Flows to and from Earth

Solar intensity and latitude play a significant role in establishing Earth's seasons and the biomes at various latitudes. The Earth's axis of rotation tilts at an angle of about 23.5° relative to the line that is perpendicular to the sun. For this reason, equatorial regions receive more direct sunlight than polar regions.

Seasons occur for reasons that are related to the Earth's tilt relative to the sun. When the Northern Hemisphere is tilted away from the sun, it experiences winter and the Southern Hemisphere (tilted toward the sun) experiences summer, and vice versa.

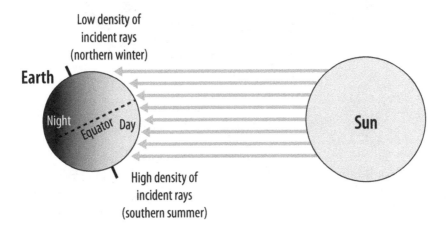

Figure 4-5. Position of the Earth and the Sun During Seasons

Atmospheric Circulation

Convection cells in the atmosphere are initiated by the warm equatorial air that rises to a higher altitude, which creates a low-pressure area underneath it. The air then moves north or south at high altitude until it confronts cool air. The cooler combined air mass becomes denser and sinks, creating a high-pressure area underneath the descending air. These high-pressure zones occur at latitudes of about 30° north and south of the equator. At these latitudes, the high pressures from the descending air produces hot, dry air. Most of the world's deserts exist at those latitudes.

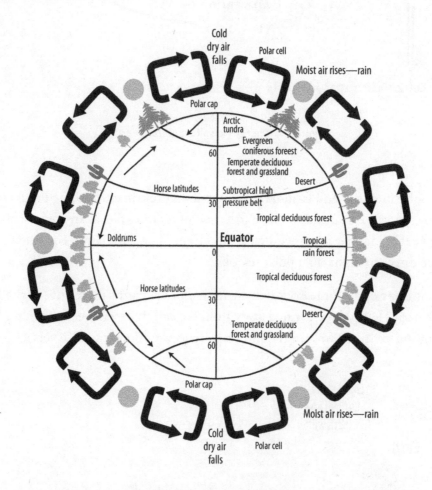

Figure 4-6. Atmospheric Circulation

The **Coriolis effect** is created by a combination of the convection cells described above, and the Earth's rotation moving underneath the cycling masses of air. The combined result is a global pattern of circular air movements. The Coriolis effect moves global wind currents in a clockwise direction in the Northern Hemisphere and a counterclockwise direction in the Southern Hemisphere. The portions of these air movements that are next to the Earth are called trade winds. The higher altitude portions of these air movements are referred to as jet streams.

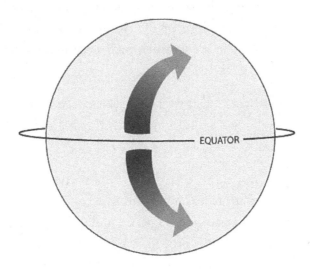

Figure 4-7. Global Air Movements

Weather

Weather takes place mostly in the troposphere. It is affected by temperature, pressure, and moisture.

The **temperature** of the air is affected by convective cycles in the atmosphere. Air of different temperatures is moved by the trade winds, driven by the Coriolis effect. Because of the high specific heat of water, air temperature varies more over land than over water.

Pressure of the air is affected by altitude and by ascending and descending air masses created by convection cells. Air pressure differences create wind on the surface of the Earth, in addition to trade winds. Air always moves from high-pressure zones to low-pressure zones.

Moisture in the atmosphere is a form of stored heat. Water vapor is also a greenhouse gas that helps hold heat energy in the atmosphere. The amount of moisture that can be held in the atmosphere increases with temperature and pressure. Therefore, a low-pressure system allows water vapor to precipitate as rain and temperature to drop. Dew point is the temperature at which condensation occurs at any give pressure.

Weather fronts are created when an air mass of one temperature and pressure collides with an air mass of different temperature and pressure.

- **Cold fronts** occur when cool air moves into a warmer area. The dense, cooler air pushes itself under the warm air. The warm air is pushed higher and is cooled, which allows water vapor in the warm air to condense and fall as rain. Cold fronts often bring severe weather, such as cumulous clouds, thunderstorms, and tornados.

- **Warm fronts** occur when warm air displaces cool air. The lower density warm air slides above the cool air, forming cirrus clouds and possibly a steady drizzle, if any rain falls at all.

Tropical cyclones develop in equatorial regions when vast amounts of water rapidly evaporate. The water vapor carries energy into the atmosphere. As a result of atmospheric convection, the water vapor rises and the water molecules condense and fall as rain. If large amounts of water go through this process, the resulting storm is extremely powerful, and has a characteristic rotation fueled by the Coriolis effect. In the Western Hemisphere, tropical cyclones are called **hurricanes**.

TEST TIP

Cyclonic storms and weather patterns have not typically been a major focus on the AP Environmental Science exam. However, as a result of global warming, an increase in the severity and frequency of tropical cyclones has been cited. If the trend continues, future APES tests may well seek an understanding between these storms and increased rates of water evaporation at the equator.

Climate

Climate is the long-term pattern of meteorological conditions in a particular area. By contrast, the term, **weather** refers to meteorological conditions on a particular day or over a very short time. (See the section Global Warming in Chapter 10.)

Ocean currents affect climate by carrying heat energy from one part of the Earth to another. The high specific heat of water means that the air tends to shift in temperature toward the temperature of the water. Warm water brings warmer weather; cool water keeps the air cooler. Consequently, the direction of ocean currents—driven by trade winds—and the time the ocean water spends at a particular latitude contribute greatly to long-term patterns of air temperature, moisture, and, therefore, overall climate.

TEST TIP

The body of material for the AP Environmental Science exam is very interdependent, just as the systems of the Earth are interdependent. As you read about global events, such as ozone depletion or the greenhouse effect, be sure to return to this chapter to see how these topics are related.

El Niño Southern Oscillation (ENSO) is a periodic shift in climate patterns based on a change in the direction of currents in the Pacific Ocean. Normally, the surface currents move equatorial waters westward from South America to Indonesia and Southeast Asia. As the water moves west, it is heated and evaporates. The trade winds carry the warm water vapor. By the time the air mass reaches Asia, it is laden with moisture, and the precipitation provides water for the tropical rainforests in that region. However, during an El Niño event, the equatorial current flows east instead of west. This brings warm water and moisture to South America rather than Asia. As a result, the cooler jet stream precipitates the moisture in Central or North America, and normally moist areas in the western Pacific are much drier.

DID YOU KNOW?

Peruvian fishermen were the first to notice El Niño events because the reversal of the water's movement prevented nutrients from upwelling off the coast of South America and sharply reduced their harvest of anchovies. The name "El Niño" refers to the infant Jesus because these times of lower fish catches took place around Christmas.

Global Water Resources

Global water resources are contained in different compartments within the hydrosphere, which includes the land, the atmosphere, and the oceans. The availability of fresh water is a key concern as humans try to engineer a predictable, sustainable plan for the future. (See Chapter 7 for a more thorough review on managing water resources.)

The **water cycle** describes the way water circulates around the Earth and is stored in different global compartments.

The water cycle is driven by solar heating and gravity. Energy from the sun evaporates water from oceans, inland bodies of water, leaves, and soil. It then moves through the air as water vapor. If the water vapor cools or decreases in pressure by moving to a higher altitude, liquid water precipitates as rain or snow. If the rain lands on terrestrial ecosystems, it can enter a river and become runoff water or it can infiltrate the ground as groundwater. The cycle begins again as water flows to the sea and again evaporates.

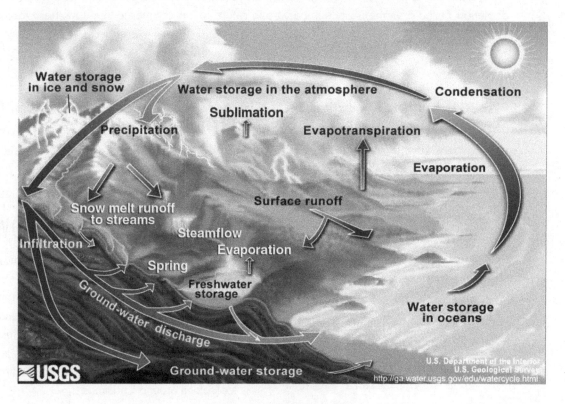

Figure 4-8. The Water Cycle

Water compartments describe the places that retain water at different points in the water cycle. The fresh water that humans need and use comes mostly from lakes, reservoirs, rivers, and groundwater. However, the combined percentage of all freshwater that is available from these sources represents less than one-third of one percent of the global supply of water. Over 97% of Earth's water is contained in oceans as salt water. (See Chapter 7 for more detailed information on global water compartments and how they are managed.)

Groundwater

As water is absorbed into the ground and moves downward, it eventually reaches an impervious layer of rock and forms an aquifer. Aquifers that are charged by percolation from above are called unconfined aquifers. The upper boundary of an unconfined aquifer is called the water table.

A confined aquifer is sandwiched between two layers of impermeable rock. Confined aquifers are charged in recharge zones where the layer of the sandwiched, porous rock absorbs water directly through a lake, river, or some other type of runoff.

> **TEST TIP**
>
> Groundwater pollution and depletion is a major issue in some parts of the United States. It is highly likely that there will be questions on the exam that ask about your knowledge of groundwater pollution.

Well water is obtained by drilling a well into the groundwater. If too many wells are drilled, the aquifer will subside, and perhaps form a dip in the water level around the well, called a cone of depression. If this occurs near the coast, where an aquifer empties into the sea from underground, it is possible for saltwater to diffuse up into the aquifer, thus spoiling the water. This saltwater spoilage of coastal wells is called saltwater intrusion. As the water from an intruded well becomes saltier, a user may be tempted to use the water for irrigation. Doing so will leave a salt film on plants and in the soil, called salinization, which will eventually kill the plants and degrade the land.

Table 4.2 Location of Water in Earth's Compartments

Water Compartment	% of Global Water Supply	Average Residence Time
Ocean	97.60	3,000–30,000 years
Ice and snow	2.07	1–16,000 years
Groundwater to 1 km	0.28	days–1,000s of years
Soil, animals, plant moisture	0.010	weeks
Lakes and reservoirs	0.009	1–100 years
Saline lakes, inland seas	0.007	10–1,000 years
Atmosphere	0.001	8–10 days
Swamps and marshes	0.003	months to years
Rivers and streams	0.0001	10–30 days

Ocean circulation is caused by one of three forces: currents induced by *global wind patterns*, thermohaline currents, and tides.

1. Trade winds and the Coriolis effect cause friction on the surface of the ocean and drag the surface water in the same direction as the wind. Consequently, the major ocean currents also move in circular paths that are clockwise in the Northern Hemisphere and counterclockwise in the Southern Hemisphere. When major currents encounter continents, they tend to split and move in both directions along the coast of the continent. Trade winds and the placement of continents dictate the direction of major surface currents in the oceans.

2. **Thermohaline currents** are caused when both temperature and/or salinity changes the density of water, which results in water moving from high-density areas to low-density areas. Thermohaline currents are responsible for most subsurface currents. When two masses of water collide, the water of lower density moves to the more shallow depth.

3. **Tides** are caused by the gravitational pull of the moon on the water. This pull creates a bulge of water around the globe that faces the moon, and another bulge of water on the opposite side of the Earth. As the moon passes around the Earth roughly once per day, these two tidal bulges also move, creating a high tide and a low tide twice a day.

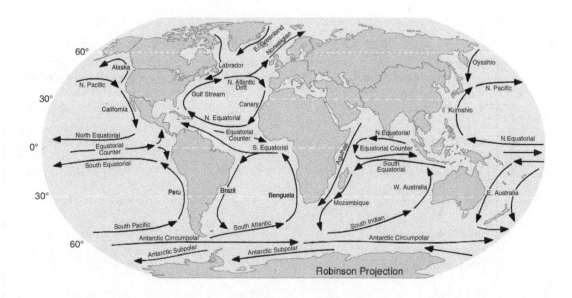

Figure 4-9. Ocean Circulation

Soil Resources

The **rock cycle** consists of the cycling of Earth material among three types of rocks: **igneous**, **sedimentary**, and **metamorphic**.

1. **Igneous rocks** are formed from magma that has solidified. Magma is the molten material that comes from the Earth's mantle during volcanic activity. Above the Earth's surface, magma is called lava. Basalt, rhyolite, and granite are types of igneous rocks.

2. **Sedimentary rocks** are formed when weathering processes break rocks into particles and deposit them in layers. Over time, these layers compress and harden to form sedimentary rock. Different types of sedimentary rock include shale, limestone, and sandstone. Weathering occurs from either mechanical or chemical forces.

    a. **Mechanical weathering** takes place when a force is applied against the rock to break it up. Glaciers, ocean waves, and erosion from wind or water are examples of mechanical weathering.

    b. **Chemical weathering** takes place when a chemical reaction weakens and breaks apart rock surfaces. For example, the oxygen in the air can react with some minerals and break them up. Acid deposition can dissolve some minerals. Chemical weathering weakens rocks so that they undergo mechanical weathering more easily.

3. **Metamorphic rocks** can come from both sedimentary and igneous rocks as a result of being exposed to heat and/or pressure. Metamorphic rocks are usually formed when layers of sedimentary or igneous rocks become further compressed by tectonic plate movement. For example, compression of limestone produces marble; compression of shale produces slate; and compression of sandstone produces quartz. In turn, metamorphic rocks can be melted and subsequently cooled to form igneous rock.

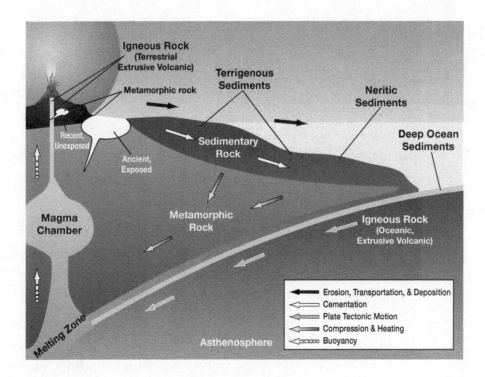

Figure 4-10. The Rock Cycle

Soil formation begins when erosion creates small particles of rock and minerals. The combination of these inorganic particles with organic material from dead plant and insect material forms soil. The process of making soil takes time. A good, fertile topsoil takes about a year to build one millimeter (mm), given good conditions.

Soil composition involves some proportion of the following materials:

1. **Clay** comes from fine particles. Clay has high adhesive ability due to mobile ionic attractions between the particles, and is impermeable to water.

2. **Silt** comes from particles that are slightly larger than clay and that result from mechanical weathering. Silt particles combine to form sedimentary rock when compressed.

3. **Sand** particles are larger than both silt and clay and come from mechanical weathering.

4. **Gravel** particles are larger than sand and come from microscopic pebbles and rock fragments.

5. **Humus** is a sticky composite of organic material made up of dead and decaying plants and animals. Humus provides the "glue" that holds together the inorganic parts of the soil and gives soil its spongy texture.

6. **Loam** is a mixture of sand, silt, humus, and clay and is well-suited for growing crops. Varying the level of sand in the loam adjusts the soil's ability to retain water.

Physical and Chemical Properties of Soil

1. **Texture** is determined by particle size—smaller particles feel smoother; larger particles feel coarser.

2. **Porosity** refers to the ability of water to flow through the soil. When soil has larger particles, there is more space between the particles, which allows water to flow through it and allows plants to develop roots. Low soil porosity means the soil is more compact.

3. **Permeability** is the ability of water to flow through the soil. Soil with low permeability retains water more easily. Soil with high permeability allows water to flow through it easily.

4. **Fertility** refers to the amount of nutrients carried by the soil. Essential nutrients include nitrogen and phosphorus in various forms. (See the section called Natural Biogeochemical Cycles in Chapter 5.)

5. **pH** is a measure of soil acidity. The normal pH range for soil is between 6.0 and 8.0.

6. **Salinity** is a measure of a soil's salt content. In most cases, a lower salinity means that the soil is more able to grow plants.

Soil horizons refer to the layers of different types of material in the soil. A cross-section of these horizons, or layers, is called a soil profile. Soils in different places on the Earth have different proportions of each horizon. The different soil horizons, starting with the layer closest to the surface, are:

1. **O-Horizon** is the organically rich surface that contains leaves and partially decomposed organic material.

2. **A-Horizon** is the fertile topsoil, which is composed of different combinations of inorganic material and organic material.

3. **B-Horizon** is the subsoil, which has lower levels of organic material and higher levels of inorganic clay and silt. Water soluble compounds from higher horizons percolate down to this horizon. This horizon can also contain weathered material from the C-Horizon just below. The B-Horizon often becomes compressed to the point that roots cannot penetrate further down.

4. **C-Horizon** contains weathered parent material—or rock—with little or no organic material. Most of the parent material in the United States was transported by glaciers and other geological events, rather than being composed of the underlying bedrock.

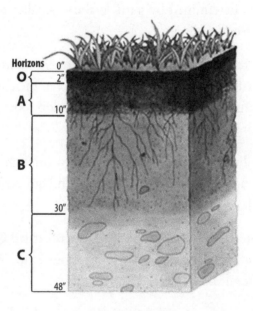

Figure 4-11. Soil Horizons

Soil erosion occurs when soil is moved from its point of origin in one of three ways:

1. **Rill erosion** occurs when water cuts small rivulets in the soil.

2. **Gully erosion** occurs when rill erosion grows to form a large channel.

3. **Sheet erosion** occurs when water removes a horizontal layer of soil.

Causes for erosion include the following:

1. **Wind and water** can remove the top layer of nutrient-laden adhesive topsoil.

2. **Chemical** erosion is caused by nutrient depletion, acid deposition, or salinization.

Soil conservation practices (more information about soil management practices can be found in Chapter 7).

1. **No-till or low-till farming**, which leaves plant material lying on the ground after harvest to protect the soil from wind and to retain moisture.

2. **Windbreak trees** can be planted to shield the soil from prevailing winds.

3. **Contour farming** or **terracing** reduces water erosion by keeping water from flowing downhill.

4. **Adding soil nutrients** so that the adhesive quality of the soil is maximized, and the soil resists movement due to erosion.

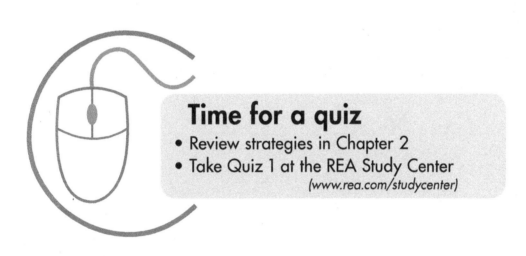

Time for a quiz
- Review strategies in Chapter 2
- Take Quiz 1 at the REA Study Center
(www.rea.com/studycenter)

Chapter 5

The Living World

This chapter describes how the living components of the environment combine with the nonliving components to make up ecosystems. It describes the relationships between living organisms and provides a survey of Earth's major ecosystems. This is the major exposure to basic biology that you will have in your AP Environmental Science course.

The College Board syllabus indicates that the following topics from this chapter may show up on the AP Environmental Science exam. These topics compose about 10–15% of the total test.

A. **Ecosystem Structure**
(Biological populations and communities, ecological niches, interactions among species, keystone species, species diversity and edge effects, major terrestrial and aquatic biomes)

B. **Energy Flow**
(Photosynthesis and cellular respiration, food webs and trophic levels, ecological pyramids)

C. **Ecosystem Diversity**
(Biodiversity, natural selection, evolution, ecosystem services)

D. **Natural Ecosystem Change**
(Climate shifts, species movement, ecological succession)

E. **Natural Biogeochemical Cycles**
(Carbon, nitrogen, phosphorus, sulfur, water, conservation of matter)

Ecosystem Structure

Ecosystem composition includes biotic (living) and abiotic (nonliving) factors. Each organism in the ecosystem lives most comfortably within abiotic factors that are defined by its tolerance limits.

- **Biotic factors** refer to living components in an ecosystem. Biotic factors include the species present, the sizes of their populations, and the relationships between those species. The biotic factors include the entire community of different species.

- **Abiotic factors** refer to nonliving components in an ecosystem. Abiotic factors include the geochemical cycles—or amounts of each chemical present, temperature, pressure, salinity, pH, and humidity.

Biological Populations and Communities

Biological populations and communities represent the key biotic component of ecosystems.

- An **organism** is a single living individual and represents the simplest free-existing unit of a biological community.

- **Populations** are composed of a group of interbreeding organisms of the same species that live in the same general area at the same time. Remember that a *species* is a group of organisms whose genetics are similar enough that, when they reproduce, they produce fertile offspring.

- A **community** is a group of interdependent populations whose niches overlap in some way, usually by geographical location.

Traits of Biological Communities

Ecological niches are the combination of the place where a particular population lives—including all the abiotic factors it requires (habitat)—and the ecological role that the species plays in the ecosystem. Ecological role refers primarily to the method of obtaining food and the relationships an organism has with other organisms in the community.

Keystone species are species that have niches upon which other species depend to a high degree. If the keystone species were to be removed, most of the other species in the community would not be able to survive.

Indicator species are species that have niches that rely on common abiotic factors with so many other species in that community that, if it can't survive, it is unlikely that most of the other species would be able to survive.

Tolerance limits are the abiotic limits that each species can tolerate. These limits define the physical limits of the niche of each organism. Tolerance limits are usually expressed as a bell curve, with the abundance of organisms expressed by the vertical axis and a range for a given abiotic factor—or environmental gradient—on the horizontal axis. The graph that results is called a tolerance curve.

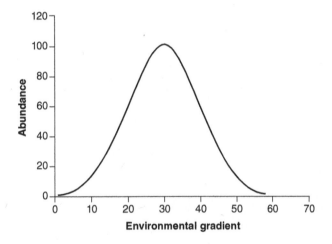

Figure 5-1. Tolerance Curve

For example, humans are comfortable living at temperatures that range from about 55° to 75°F. On both ends of that zone are temperature zones where we can exist, but we have to adjust our clothing, cooling, or something so that we can exist for an extended length of time. On the extreme ends of these zones are our tolerance limits, let's say about −10°F and 115°F, where we can exist for a short time but could not live indefinitely.

TEST TIP

While the concept of tolerance limits is not directly mentioned in the AP Environmental Science syllabus, knowing about them may help you give an exceptionally strong answer to free-response questions that ask you to connect how living things respond and thrive to physical parameters in the environment.

Interactions among species may include predation, competition, or symbiotic behaviors that help species survive and reproduce.

Symbiosis is a relationship between two organisms that co-evolve together. The three major types of symbiosis are **parasitism, commensalism,** and **mutualism**.

- Parasitism is when one organism derives all of its nutrients from a host organism. A parasite is more successful if it does not kill its host. Ectoparasites infect the outside of a host; endoparasites live inside a host.

- Commensalism is the type of symbiosis in which one organism benefits and the other is neither benefited nor harmed.

- Mutualism is the type of symbiosis in which both organisms in the relationship benefit.

Competition occurs when the niches of organisms sometimes overlap. Two individuals need the same resources to survive. The greater the niche overlap, the greater the level of competition. Eventually one or both species will engage in resource partitioning or adapt a new niche to reduce competition.

- **Intraspecific competition** occurs when individuals of the same species compete for resources. Intraspecific competition shows up in the population distribution of a species. An *ordered* distribution suggests that members of the population compete for resources. A *clustered* distribution suggests that members of a population receive some mutual benefit from being close; thus, competition is low. A *random* distribution suggests that resources are readily available, and no cooperation or competition is present.

- **Interspecific competition** occurs when individuals from different species compete for resources. For example, rainforest plants compete for available sunlight under a dense canopy by evolving large leaves. Interspecific competition pushes other species to adapt more quickly than other species through the process of natural selection.

Survey of Earth's Ecosystems

Earth's terrestrial biomes and aquatic ecosystems are covered in the two charts in this chapter.

> **TEST TIP**
>
> The chart-style presentation of terrestrial biomes and aquatic ecosystems will help you when you answer compare/contrast questions about the different biomes. Practice by comparing locations of these biomes on a map or comparing types of plant adaptations in biomes of different climates.

Terrestrial Biomes

Terrestrial biomes are most influenced by climate (especially moisture and temperature), which is dictated by the position of the biome on the Earth. Each biome is most noted by the adaptations that organisms have to these abiotic factors.

Table 5-1. Terrestrial Biomes

Biome	Rainfall and Climate	Key Producers	Producer Adaptations	Notable Consumers	Location
Tundra	Less than 10 inches of rain per year. Extreme cold.	Lichens, low-lying shrubs.	Cold climate freezes water before it seeps into ground. Little top soil develops. Plants adapt to short growing season, moisture retention.	Low species complexity, constancy, inertia, and renewal abilities. Consumers mostly primary.	High latitudes near poles (arctic tundra) and high altitudes (alpine tundra).
Desert	Less than 15 inches of rain per year. Very dry.	Low dispersed shrubs in rocky, sandy soil. Cactus, yucca.	Low moisture levels prevent topsoil from accumulating. Moisture comes in periodic thundershowers. Plants must quickly absorb and retain water; waxy, small leaves.	Animals tend to be smaller, remain underground during the day, and seek other methods to retain moisture and protect from heat.	Typically at high-pressure zones on the leeward side of mountain ranges, or at 30° north or south latitude.

(continued)

(continued)

Biome	Rainfall and Climate	Key Producers	Producer Adaptations	Notable Consumers	Location
Grassland/ Savannah	10–35 inches of rainfall per year for grassland; more for savannahs. Short growing season. Humid summers.	Grasses, frequently turned into agricultural areas.	Plants tend to be fire-adapted in some way. Fires make it difficult for trees to grow. Desertification a risk if soil not tended well.	Coyotes, buffalo, prong-horn.	The Great Plains of the United States, Russia, and Argentina. The term *savannah* refers to grasslands in tropical regions, such as Africa.
Chaparral	10–30 inches of rainfall per year. Dry.	Mesquite, cactus, manzanita.	Plants adapted to fire and moisture conservation. Plants shed leaves during dry season.	Snakes, coyotes	California, Mediterranean.
Temperate deciduous forest	30–60 inches of rainfall per year. Four seasons.	Hardwood trees such as maple and oak.	Fertile soils from decomposed leaves create opportunities for decomposers, mosses, and underbrush.	Consumer/decomposer of the largest biomass is the earthworm. Macrofauna include squirrels, bear, birds, deer.	Central Europe and Eastern United States.
Taiga or boreal forests	15–35 inches of rainfall per year. Dry, cold winters; wet summers.	Lodgepole pine, spruce, fir.	Evergreen needles maximize photosynthesis from sparse sun and inhibit loss of moisture.	Prominent consumers include bear, elk, deer, moose, eagles.	High latitudes just under the arctic tundra. Canada, northern United States, Alaska, and Russia.
Tropical and temperate rainforests	Two seasons: rainy and dry.	Tall deciduous canopy, with epiphytes growing among the trees.	Broad leaves in plants to outcompete other plants for sun. Broad root structures to support tall trees in infertile, shallow soil. In this highly diverse and competitive environment, nutrients are used before they reach the ground.	Wide diversity of invertebrates, birds, snakes, mammals.	Tropical forests are found within 30° north or south of the equator. Temperate rainforests tend to be found near oceans in very moist climates.
Coastal: supratidal and barrier islands	Temperature is moderated by proximity to ocean and also depends on ocean currents and latitude.	Sparse grasses with loose network of roots.	Rocky or sandy substrate. Waxy leaves limit water loss due to salt spray, sun, wind. Pioneer species on barrier islands.	Sand fleas feed on microbes, crabs feed on fleas and small sand shrimp. Birds feed on crabs.	Coasts throughout the globe. Barrier islands form on eastern continental shelf from sand transported by waves.

Aquatic Biomes

Aquatic biomes are freshwater and ocean ecosystems. Ocean ecosystems are divided into pelagic, or open ocean, and benthic, or ocean bottom, ecosystems. Freshwater ecosystems, and those influenced by freshwater, include estuaries, salt marshes, rivers, and lakes.

While terrestrial ecosystems are often characterized by the amount of moisture and type of climate, aquatic ecosystems are characterized by the amount of sunlight, nutrients, and dissolved oxygen in the water. Areas where currents have churned up nutrients that have fallen to lower levels, or deposited by a river, tend to have greater productivity. Colder water can dissolve more oxygen, but warmer water allows chemical reactions to proceed more rapidly and fosters richer plankton growth.

Table 5-2. Aquatic Biomes

Ecosystem	Depth, Sunlight Penetration	Other Key Biotic and Abiotic Factors	Notable Producers	Notable Consumers	Zones or Variations Within Ecosystem
Pelagic/open ocean	Varies in different zones.	Entirely seawater/salt water	Phytoplankton are key producers in the neritic and photic zones. Brown algae in depths of 15–60 feet; red algae in deeper water.	Zooplankton are microscopic larvae of larger animals that live in deeper water, are primary consumers and prey for small fish, which are then eaten by larger fish.	**Neritic zone:** continental shelf. **Photic zone:** open ocean where light penetrates. **Aphotic zone:** open ocean where light does not penetrate.
Benthic; coral reef	Shallow; must receive sunlight for algae within coral to grow.	Saltwater; warm climates. Coral is made up of a symbiotic relationship between algae and coral polyps.	Dinoflagellate algae is the key producer and maintains a mutualistic relationship with coral, an animal. Coral cannot live without algae—which dies during bleaching of the coral.	Coral polyps rely upon the photosynthetic algae. Also, other invertebrates live among the coral and feed off of the plankton in the highly productive coral system.	**Fore reef** receives the most wave action. The **reef crest** is the most shallow portion of the reef. The **reef lagoon** is the still portion that is most protected from oceanwave action.
Benthic; abyss	Depths over 4,000/m. No sunlight penetration; prevents photosynthesis.	Low dissolved oxygen. Many decomposers and chemoautotrophs.	Much of the energy for consumers comes from decomposers, rather than photoautotrophs, because of a lack of sunlight. However, chemoautotrophs obtain energy directly from sulfurous thermal vents.	Primary consumers may include invertebrates that are rarely seen on the ocean's surface, such as oversized clams and tubeworms. Top carnivores include giant squid. Even some whales, such as the sperm whale, dive to these depths to feed.	Below the abyss in the deepest trenches is the **hadal zone**, where there are very few nutrients at all. Above the abyss is the **bathyal zone**.

(continued)

(continued)

Ecosystem	Depth, Sunlight Penetration	Other Key Biotic and Abiotic Factors	Notable Producers	Notable Consumers	Zones or Variations Within Ecosystem
Benthic; intertidal (littoral)	Maximum sunlight.	Exposed to both seawater and atmosphere. Must adapt to wave action.	Green algae and phytoplankton.	Vast diversity of invertebrates.	Different intertidal ecosystems include **sandy beaches**, **rocky beaches**, and **mangrove swamps**.
Estuaries, salt marshes	Maximum sunlight	Mixed saltwater and freshwater; also influenced by tides.	Shallow waters are a nursery for animals, and a haven for phytoplankton. Decomposers feed from the vast nutrients flowing from the river and from sea grasses.	Consumers can come from other ecosystems and feed in estuaries, thereby transporting energy and nutrients to other terrestrial and aquatic ecosystems.	Mouths of rivers, where freshwater meets ocean water. Different estuaries characterized by varied salinity levels.
Ponds, lakes (lentic or still)	Maximum sunlight in littoral and limnetic zones. Less in profundal zone; less still in benthic zone; but all zones likely receive some sunlight except in very large lakes.	Freshwater.	**Oligotrophic** lakes are clear, with little algae and nutrients; **eutrophic** lakes are cloudier, more algae blooms, more decomposing material on bottom.	Major consumers are fish, such as trout. Terrestrial animals can also be large consumers.	**Littoral:** along the shore. **Limnetic:** surface of water in middle. **Profundal:** open water in middle. **Benthic:** Bottom.
Streams, rivers (lotic or flowing)	Maximum sunlight.	Freshwater.	Phytoplankton.	Most consumers in many lotic zones are insects in larval stages. Can also include other invertebrates, such as mollusks. Fish are top aquatic carnivores, but terrestrial vertebrates may also get food from rivers.	**Channel** is where water flows through, and may include exposed sandy areas. **Riparian zone** is the region with terrestrial organisms that transition between the channel and upland terrestrial zone.

Energy Flow in Ecosystems

Productivity

Primary productivity is the productivity of producers or autotrophs. Different ecosystems have widely diverse primary productivity. A tropical rainforest, for example, is able to convert massive amounts of carbon dioxide to plant carbohydrates because there is a lot of water available. Conversely, a desert has less water and less primary productivity.

- **Gross primary productivity (GPP)** is the amount of biomass produced by photosynthetic organisms. It is a measure of how much solar energy is converted into chemical energy through photosynthesis.

- **Net primary productivity (NPP)** is the amount of energy that is available to consumers of photosynthetic organisms. The difference between gross primary productivity and net primary productivity is accounted for by the amount of energy consumers use through respiration.

TEST TIP

The difference between primary and net productivity is an example of the Second Law of Thermodynamics, where some energy is lost as waste heat, called entropy, during any transfer of energy. As energy is transferred from the sun to the chemical bonds in plants to consumers, some energy is lost as entropy. This is an important connection to make and usually shows up in at least one place on every AP Environmental Science exam.

Secondary productivity is the amount of biomass produced by organisms that eat photosynthetic organisms. It can also be indirectly measured by the amount of waste that a consumer produces.

TEST TIP

A good exercise to prepare for the exam might be to design an experiment to measure gross primary productivity and secondary productivity.

Food Chains

Trophic levels describe the steps in a food chain. The first step on the food chain, or first trophic level, is always a **producer** that converts solar or chemical energy into more complex molecules that will, in turn, be ingested by consumers at the second trophic level. The second trophic level contains an herbivore, an omnivore, or a decomposer. The third trophic level is a carnivore or decomposer. There may be several steps on a food chain, or just a few. **Autotrophs** is another name for producers.

- **Photoautotrophs** is another name for photosynthetic organisms, such as plants.

- **Chemoautotrophs** describe those producers that use chemicals, and not the sun, as their primary source of energy. For example, some deep-sea thermal vents contain organisms that live off the chemicals that come from the vents and never see the sun.

- **Heterotrophs** are those organisms that derive energy from producers, such as herbivores (plant eating), omnivores (plant or animal eating), carnivores (animal eating), and decomposers. The last step is always a decomposer that recycles the nutrients stored in living tissue back into the inorganic nutrient cycles.

Food webs are an interrelated collection of many food chains. Each organism is connected to one or more predator or prey relationships with other organisms. *Complexity* refers to the degree that members of a food web are interconnected. A food web that is highly interconnected, with several species at each trophic level, is said to have a high level of complexity.

Energy conversions at each step in a food chain obey the First Law of Thermodynamics (energy is conserved) and also the Second Law of Thermodynamics (some heat is lost to entropy). The amount of biomass, which is proportional to the energy at each trophic level, can be visualized with a biomass, or energy pyramid, as shown in the figure below. Notice that each trophic level contains about 10% of the biomass held in the previous trophic level. The other 90% is lost to heat, metabolism, undigested food, or unconsumed food at each level of the pyramid.

> **DID YOU KNOW?**
> The largest animal on Earth, the mighty blue whale, must sustain itself on phyto- and zooplankton. By drawing directly from the producer level of the ecological pyramid, this macro-sized critter is able to find enough energy within the ecosystem to sustain itself.

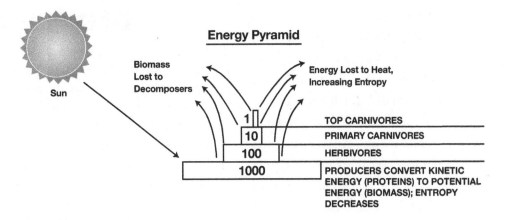

Figure 5-2. Biomass Pyramid—Three trophic levels, with energy that does not go into biomass in the next trophic level going into the environment as heat.

TEST TIP

Be able to account for the size of a food pyramid based on the Second Law of Thermodynamics. There are two reasons why each subsequent trophic level contains a smaller amount of energy: (1) some of the energy goes into metabolism, and (2) some of the energy is lost as waste heat, or entropy, as a result of inefficient energy transfers.

Ecosystem Diversity and Change

The array of species that make up an ecosystem can change for a number of reasons, including natural selection, climate shifts or some other disruptive event, species movement, or ecological succession. Each of these has a different impact on the diversity, complexity, and stability of the ecosystem.

Natural selection is the process of evolution by which members of a population with traits that survive in a particular setting live long enough to produce offspring that carry the same traits.

- The **genetic basis of natural selection** rests in the DNA that is passed on to the next generation. If the DNA produces proteins to help that organism survive and reproduce in the setting where the next generation lives, that organism will survive long enough to give that DNA to the next generation. Changes in the DNA occur through *mutation*, which is usually a harmful change in genetic material. However, sometimes mutation introduces new traits into a population that allows an organism to survive better in that organism's particular setting.

> **DID YOU KNOW?**
>
> Biomagnification was a central idea in Rachel Carson's seminal book, *Silent Spring*, which helped to usher in the modern environmental movements. Ms. Carson, a marine biologist, identified the danger of an insecticide, DDT, which was building up in the food chain. Since the United States eliminated DDT, many severely affected and consequently endangered animals—including the American bald eagle—have made enormous strides in returning to previous numbers.

- **Adaptation** is where a species evolves or otherwise develops a trait that increases its ability to survive and reproduce.

- **Resistance** is an evolutionary adaptation that helps a species survive a particular threat, such as toxicity (as in the case of bacteria becoming resistant to antibiotics) or disease (as is the case when apples become resistant to diseases like scale).

- **Speciation** occurs when two subpopulations experience different mutations or selective pressures so that they change to the point where one subpopulation cannot successfully breed with another subpopulation.

- **Divergent evolution** occurs when organisms with similar traits evolve so that they become less similar, often to minimize competition if they are in an overlapping niche.

- **Convergent evolution** occurs when different species evolve structures that are similar to one another, or occupy a similar niche in different ecosystems. For example, many types of fish tend to have lighter underbellies that make them less visible from underneath, and coloration on top that blends with the substrate below them.

Climate shift is a major abiotic change for living things. To survive a climate shift that exceeds a species' tolerance limits, that species must either adapt or migrate.

Ecological succession

1. **Primary succession** occurs when species move into an unoccupied area. Such species are called *pioneer species*. Organisms that are typically successful during primary succession reproduce quickly, survive adverse conditions, and disperse

easily. The clearest example of primary succession begins with parent material rock. Lichens begin to colonize and slowly begin to build soil. As the soil develops, a larger plant might be able to colonize the area, and so forth.

2. **Secondary succession** occurs when some change initiates a new biological community. This might occur after some disruption, such as a flood or a fire. Beavers are major agents of secondary succession in northern taiga forests, where they dam a stream, which builds a lake. The trees in the area die because their roots are underwater. The lake eventually fills in and becomes a meadow, which eventually undergoes succession until it becomes a forest again.

3. **Mature communities** exist when succession has proceeded to the point where the biological community resists further change in composition. Different ecosystems are characterized by the climax community that they support. For example, the Yellowstone taiga forests seem to culminate with a pine forest until thunderstorm-ignited fires initiate secondary succession once again.

Biodiversity

Biodiversity occurs at one of three levels: within the genes of a single population, among species within a single community, or within the number of ecosystems on a global scale.

Genetic biodiversity refers to the diversity of genes that show up within a population of a single species.

- **Genes** are portions of DNA that contain the information needed to produce specific proteins.

- **Gene expression** is the cellular process of converting the information within the gene into a protein.

- **Gene pool** is the collective group of traits that exist in all the chromosomes of all the individuals in a population.

- **Founder effect** is a limit to genetic diversity that is created when a small group of organisms begins a new population.

> **DID YOU KNOW?**
> The bacteria that are responsible for the brilliant colors in Yellowstone National Park's hot springs have evolved proteins that replicate DNA at very high temperatures. When scientists were looking for ways to replicate DNA at very high temperatures in the laboratory, they found the natural proteins they needed in these hot springs bacteria!

- **Genetic isolation** is one aspect of the founder effect, in which a small number of individuals have been isolated. The number of genes available from the original breeding pair limits the gene pool of the resulting population.

Species biodiversity refers to the diversity of species in a community within a single ecosystem. This is sometimes called *species richness*. For example, there is greater species diversity in a tropical rainforest than there is in an arctic tundra.

- **Complexity** is the number of species at each trophic level.
- **Abundance** is the number of individuals in a species.

Ecosystem biodiversity refers to the diversity of different ecosystems globally. For example, if all coral reefs were to be destroyed due to global warming, the overall global diversity of ecosystems would decrease.

Natural Biogeochemical Cycles

Biogeochemical cycles, like the water cycle discussed in Chapter 4, are the ways in which matter is conserved on Earth. Biogeochemical cycles include the carbon cycle, the nitrogen cycle, the phosphorous cycle, and the sulfur cycle.

The Carbon Cycle

The **carbon cycle** is involved in converting the sun's energy into important molecules for life (photosynthesis), the process of metabolizing molecules (respiration), as well as climate change. The carbon cycle involves the following components:

1. **Photosynthesis** occurs when plants absorb carbon dioxide and water in the presence of sunshine, and produce oxygen and carbohydrate molecules.

2. **Respiration** occurs in animals as they metabolize carbohydrate molecules in the presence of oxygen, and produce carbon dioxide and water.

3. **Carbon sequestration** is the storage of carbon in the following forms. Each of these forms is like a reservoir of carbon. Releasing carbon from any of these forms into the atmosphere accelerates climate change, because carbon dioxide is a greenhouse gas. (See Global Warming in Chapter 10.)

- **Fossil fuels** are formed when bacteria, plants, and animals die and decompose. During this decomposition, hydrogen atoms attach to the carbon from the proteins, carbohydrates, and nucleic acids in their bodies to form hydrocarbons. Crude oil and natural gas are a mixture of hydrocarbons that comes from decomposed bacteria. Coal is mostly carbon and comes from decomposed plants. Carbon from these sources can reenter the carbon cycle when these fuels are burned (combusted) and carbon dioxide is produced.

- **Carbonate ions** dissolved in water are in equilibrium with carbon dioxide in the atmosphere. Carbonate ions are picked up by some marine creatures to build calcium carbonate exoskeletons from the microscopic level (diatoms) to the macroscopic level (coral reefs).

- **Limestone** is a sedimentary rock that is formed when diatoms, corals, and other calcium-containing organisms die and leave their calcium carbonate exoskeletons.

- **Polar ice** entraps methane within it. Some polar ice contains so much methane that it can actually burn!

- **Forests** are considered a type of carbon sequestration that actively removes carbon dioxide from the atmosphere and stores it within the chemical bonds of its biomass.

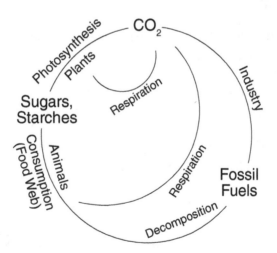

Figure 5-3. The Carbon Cycle

> **TEST TIP**
>
> Familiarity with the carbon and nitrogen cycles is critically important for the AP Environmental Science exam. You should understand every aspect of them. In fact, some aspect of the carbon cycle frequently shows up on a free-response question. Although the phosphorous and sulfur cycles are less likely to take an important role on the exam, you should understand the basic concepts of these cycles, too.

The Nitrogen Cycle

The **nitrogen cycle** is summarized in the figure below and is made up of the following major steps:

1. **Nitrogen fixation** occurs when soil bacteria and bacteria in legume root nodules turn nitrogen gas in the atmosphere into ammonia. Ammonia is a critical nutrient for many plants. Lightning is capable of oxidizing nitrogen gas in the atmosphere directly into oxides of nitrogen; this process is also considered to be nitrogen fixation.

2. **Nitrification** occurs when bacteria oxidizes ammonia into nitrite and nitrate ions.

3. **Denitrification** occurs when bacteria turns oxidized nitrogen (nitrates and nitrite ions) into nitrogen gas.

4. **Ammonification** occurs when decomposers convert animal and plant protein into ammonia.

5. **Assimilation** occurs when plants absorb ammonia and nitrate ions to build large, essential molecules, such as proteins and DNA.

> **DID YOU KNOW?**
>
> Sewage treatment plants are an application of the nitrogen cycle. Composting kitchen waste is an application of both the nitrogen and carbon cycles!

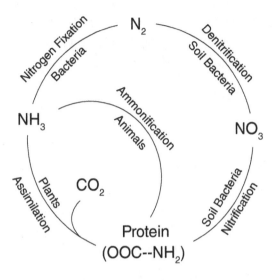

Figure 5-4. The Nitrogen Cycle

The Phosphorus Cycle

Plants and animals use the phosphate ion in essential chemical reactions, particularly in respiration and photosynthesis. Decay puts phosphates into the soil where it is later mined and used as fertilizer, or it becomes a part of rocks in geological processes. Plants later pick up the phosphate ions from the soil, and animals ingest the plants.

The Sulfur Cycle

Sulfate ions are taken in by plants as essential nutrients to produce proteins. Disulfide bonds in the proteins of living things are critical in maintaining the correct shape of each protein molecule.

Fossil fuels contain sulfur because the disulfide bonds in proteins decompose when living things die.

Sulfur oxides is released from thermal vents and volcanic activity, and also when fossil fuels are burned. When sulfur dioxide comes into contact with water vapor, it forms acid rain.

Sulfide ions are released by volcanic activity and form insoluble compounds with metal ions, or they are used by undersea chemoautotrophs as a primary source of energy. Some microbes can reduce sulfide ions into elemental sulfur.

Populations

Chapter 6

Population growth represents a fundamental cause of environmental challenges, such as pollution and the increased demand for resources. At the current growth rate of about 1.1%, the Earth's population will double in about 64 years to over 14 billion people. This poses a major challenge as we plan for a secure and sustainable future. Therefore, your careful attention to reviewing these topics is central to doing well on the AP Environmental Science exam.

The College Board syllabus indicates that the following topics from this chapter may show up on the AP Environmental Science exam. These topics compose about 10% to 15% of the total test.

A. **Population Biology**
   (Concepts: population ecology, carrying capacity, reproductive strategies, survivorship)

B. **Human Populations**
   1. Human population dynamics (historical population sizes, distribution, fertility rates, growth rates and doubling times, demographic transition, age-structure diagrams)
   2. Population size (strategies for sustainability, case studies, national policies)
   3. Impacts of population growth (hunger, disease, economic effects, resource use, habitat destruction)

TEST TIP

This chapter includes a number of figures and graphs that succinctly summarize a lot of information. Don't pass over them lightly—be sure that you understand and can interpret each graph. It will help you come exam day.

Population Biology

Key Terms

1. **Fertility** is a measure of the actual number of offspring produced. It is often expressed statistically as the **crude birth rate**, which is the number of individuals born per thousand people per year.

2. **Fecundity** is the physical ability of an organism to reproduce. Fecundity is what people actually refer to when saying "fertility" in casual conversation.

3. **Natality** is the production of new individuals.

4. **Morbidity** is the level of illness in a population.

5. **Mortality** is a measure of the actual number of individuals who die in a population. It is often expressed statistically as the **crude death rate**, which is the number of individuals who die per thousand per year.

6. **Survivorship** represents the number of people in a given age bracket who continue to remain alive each year.

7. **Life expectancy** is the most probable number of years an individual will survive.

8. **Lifespan** is the longest length of life reached by a given species.

9. **Total growth rate** is the sum of the increases to the population due to immigration and births, minus those individuals who have died or emigrated away.

10. **Natural growth rate** is the population growth due to births and deaths, usually expressed as the crude birth rate minus the crude death rate.

Growth Rates

Calculating growth rates is an easy process of addition and subtraction. One only needs to be aware of the factors that make up total population growth.

$$\text{Total population growth} = B - D + I - E$$

Where B = Birth rate, number born per 1,000 people in one year

D = Death rate, number died per 1,000 people in one year

I = Immigration growth per 1,000 people in one year

E = Migration away per 1,000 people in one year

EXAMPLE A town of 1,000 people experiences 16 births and 12 deaths. What is the total population growth as a percentage? (Assume no immigration or emigration.)

SOLUTION Using the above formula, $B = 16$, $D = 12$ (per 1000). Then multiply by 100 to get a percentage.

$$\frac{16 - 12}{1000} \times 100 = 0.40\%$$

EXAMPLE A town of 20,000 people experienced a crude birth rate of 48 and a crude death rate of 18. Crude immigration and emigration rates were 12 and 3, respectively. What is the total annual rate of growth as a percentage?

SOLUTION Using the above formula, $B = 48$, $D = 18$, $I = 12$, and $E = 3$. Then multiply by 100 to get a percentage.

$$\frac{48 - 18 + 12 - 3}{1000} \times 100 = 3.9\%$$

TEST TIP

Growth rate questions tend to be easy to answer if you are aware of the relationships stated above, but they are difficult to interpret if you are not well versed in these relationships. Build your test confidence by understanding what goes into calculating a total growth rate for a population.

Exponential growth is the typical pattern for population growth because each new generation of individuals produces more potential reproducers than the previous generation. In each year, the amount of population increase is, itself, increasing. Exponential growth is contrasted with arithmetic growth, which is linear and increases by the same amount each year. Exponential growth is most easily described by an annual percent growth, or doubling times.

The graph that depicts exponential growth is shaped more like a J. The maximum rate of growth that a population can experience without any resistance is called the *biotic potential*.

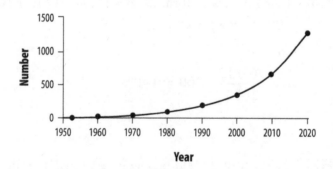

Figure 6-1. Exponential Growth By Decade

1. **Doubling time** is the time it takes for a population to double in size and can be easily calculated without a calculator using the Rule of 70:

    Annual percent growth × doubling time = 70

EXAMPLE A country grows at an annual rate of 4%. How many years will it take for the country to have a population that is four times as large?

SOLUTION For the population to increase by a factor of four, it must double twice. Find the doubling time and then double it to get the answer.

$$\text{Doubling time} = \frac{70}{4} = 17.5 \text{ years}$$

The population will quadruple after 35 years.

EXAMPLE Calculate the annual percent growth of a population that doubles in size every 10 years.

SOLUTION Annual percent growth = $\frac{70}{10}$ = 7%

TEST TIP

Exponential growth problems are typically answered with more advanced math skills. Remember, you won't have a calculator to use while taking the APES exam. The Rule of 70 has been enough to answer correctly the exponential growth problems found on past AP tests. Know the "Rule of 70" and you'll be all set to answer the exponential growth problems on your test.

2. **Logistic growth** occurs in populations that experience some *environmental resistance*, or downward pressure on the J-curve of exponential growth. In fact, most populations do not reach their biotic potential without some resistance. Population growth slows as the resistance increases. This type of growth is called logistic growth, and the growth curve looks more like an S.

Carrying capacity is the maximum population size that can be reached as a result of environmental resistance. It is represented by the top of the S on an S-shaped logistic growth curve.

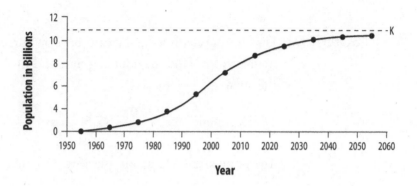

Figure 6-2. Logistic Growth By Decade

Factors That Affect Population Size

Density dependent factors tend to be biotic factors that regulate population growth.

Interspecific factors come into play when two species have a symbiotic relationship, such as mutualism, commensalism, or parasitism. In this situation, the growth curve of one population will be related and dependent upon the other. For example, in the following figure, the wolf population size cycles with the rabbit population size because the wolf preys upon the rabbit. Lots of rabbits bring lots of wolves, which in turn decreases the rabbits, which in turn decreases the wolves.

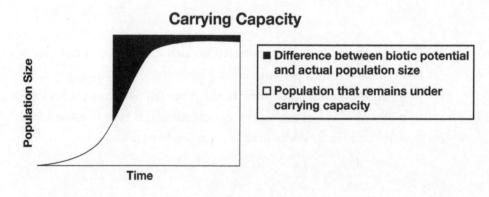

Figure 6.3 Comparison Between the Biotic Potential and the Carrying Capacity of a Population

Intraspecific factors occur when individuals of a population compete with one another, which suggests that there is a limiting resource, or environmental resistance. For example, a regular distribution of an organism—whether it be penguin nests or clumps of grass—suggests that there is competition for a resource. The grass might be competing for water, for example, and the species will have a better chance for survival if the grass is evenly distributed.

Density independent factors tend to be abiotic factors that regulate population growth. Weather and climate tend to be the most prevalent density-independent factors. Volcanic eruptions, severe storms, and fire all occur without regard to the size of the population, but they can all affect the size of a population.

Immigration and emigration change the size of a population through migration, although they do not change the size of the overall world population.

Pronatalist pressures are those aspects of the environment or culture that increase the desire to have children, which may include the following:

- **Social factors**, such as companionship, community, or religious status.
- **Financial factors,** such as a labor force on a farm, or social security in old age.
- **Fertility factors**, such as seeking "replacement" children where there is high infant mortality.
- **Cultural factors**, such as the need to produce an heir.

Birth reduction pressures cause people to have fewer children. Some of these pressures include:

- **Personal freedom for women** to find purpose in areas other than having children.
- **Materialism** as a motivator to spend resources on personal gratification rather than sharing with the next generation.
- **Socioeconomic status** associated with wealth may encourage couples to have fewer children.
- **Educational opportunities** that compete for time with childbearing years. As the childbearing years are postponed, the overall generation time increases, which decreases a population's overall fertility.
- **Perceived security** of the parents may cause them to rely less on producing children who would take care of them in old age.

Reproductive strategies are related to the type of population growth and age structure diagram shown by a population. Populations can be characterized by one of two different types of reproductive strategies, r-strategists and k-strategists. A population need not only show one or the other strategies; it might show one strategy under one set of conditions, and the other strategy under different conditions.

Table 6-1. r-Strategist vs. k-Strategist

r-Strategist Species	k-Strategist Species
Tend to show irruptive growth	Tend to show logistic growth
Mature quickly, short generations	Mature slowly, long generations
Tend to have short lives	Tend to have long lives
Do not care for their young	Young get more care, time, resources
High juvenile mortality	Low juvenile mortality
Tend to have many offspring	Tend to have very few offspring
Not sensitive to environmental resistance	Sensitive to environmental resistance
Tend to overshoot/die back when reaching carrying capacity; population oscillates around carrying capacity	Population growth slows as it reaches carrying capacity; reaches zero growth just underneath carrying capacity

TEST TIP

Be sure you have a few examples in mind for r-strategists and k-strategists. Compare mosquitoes and elephants, for example. Also, apply r-strategist and k-strategist concepts to the idea of human demographic transitions. Even though humans are k-strategists, are there some parallels when looking at human populations at different stages of a demographic transition?

Survivorship curves represent another way to characterize populations. As with r-strategists and k-strategists, no single generalization always applies to a population. In the figure below, there are four different survivorship curves that correspond to the following types of populations:

- **Population A** has low infant mortality and a high level of survivorship of most adults, such as might be the case in a developed country. There is more investment of the parents into the survival of each child.

- **Population B** has a steady mortality throughout life, independent of age.

- **Population C** has a medium level of infant mortality combined with a high level of survivorship once people have made it through childhood.

- **Population D** has a high level of infant mortality, which might be the case of an r-strategist population or developing countries.

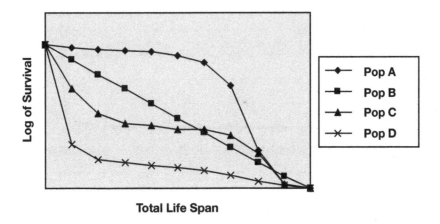

Figure 6-4. Comparison of the Survivorship Curves in Four Populations.

Human Populations

Human Population Dynamics

Historical population sizes and the impact on energy use ranging from the prehistoric to current populations are described in Chapter 8. Overall, the history of human populations can be summarized on the following graph. Even though humans are generally thought to be k-strategists, what type of strategy does this graph suggest?

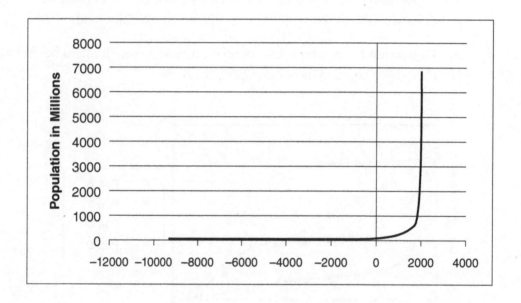

Figure 6-5. World Population Growth

> **DIDYOUKNOW?**
> Communication is one of the key skills humans used in order to build a social structure and become skillful hunters. This allowed them to expand their skills as consumers and be more ecologically successful.

Distribution of the global human population is summarized in the following world map. Currently, both China and India have populations greater than 1 billion people.

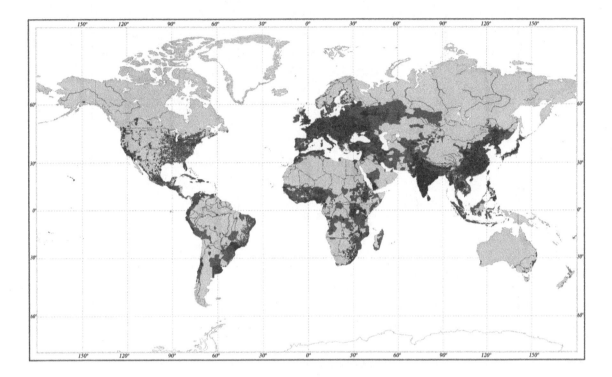

Figure 6-6. Population Density

Age-structure diagrams are histograms that depict the distribution of people at different ages within the population. These diagrams identify how many people are about to enter their reproductive years. The following set of age-structure diagrams show three different populations. The population depicted on the left shows an r-strategist population with high infant mortality. The population depicted on the right shows a k-strategist population that has reached zero-population growth. The process of evolving from a population characterized by the diagram on the left to the diagram on the right is called a *demographic transition*.

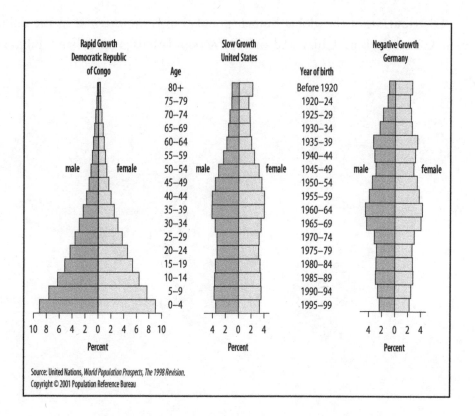

Figure 6-7. Three Patterns of Population Change (2000)

TEST TIP

The concept of a demographic transition is one of the most important concepts in understanding the relationship between culture and environmental impact. Be sure you understand the relationship among survivorship curves, reproductive strategies, age-structure diagrams, and the figure above as they describe different aspects of a demographic transition.

Demographic Transitions

Demographic transition is summarized by the following four stages, which are shown in the figure that follows. A demographic transition occurs as countries develop and population growth stabilizes. Reaching stage four of a demographic transition is critical if a culture is to become sustainable over time.

1. **Stage one** is characterized by high birth and death rates, which likely coincide with the following: little access to birth control, high infant mortality, children used as "social security" for old age, poor medical support, and famine. The age structure diagram that corresponds to this stage looks like a pyramid, with many more individuals in the younger ages. This suggests that life expectancy is not high.

2. **Stage two** is characterized by high birth rates, but death rates begin to drop—likely from improved medical care, sanitation, water quality, or food production. This results in a sharp increase in overall population.

3. **Stage three** is characterized by falling birth rates while death rates continue to fall. Population growth becomes arithmetic rather than exponential. Behaviors that may be occurring at this stage include increased contraception, less dependence by parents on children for personal security, an increase in wealth, a change in attitude toward women (careers become a socially acceptable option for women), and education.

4. **Stage four** is characterized by low birth and death rates, resulting in a steady population size. The age-structure diagram that corresponds to this kind of population has a narrow base, perhaps even rectangular. If birth rates continue to drop below death rates, the age structure diagram takes an "inverted pyramid" shape.

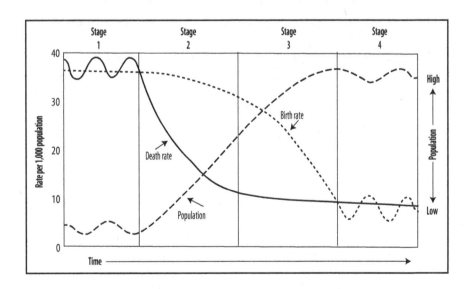

Figure 6-8. Demographic Transitions

TEST TIP

The concept of demographic transition is almost sure to show up in either the multiple-choice or free-response sections of the AP Environmental Science exam. Be sure to understand how to interpret the demographic transition figure shown on the previous page, and perhaps even duplicate it as a way to answer a free-response question.

Managing Population Size

Strategies for sustainability vary from country to country as developing and developed countries struggle to contain population growth. It seems to be clear that to manage out-of-control population growth, countries must catalyze a demographic transition. Possible strategies for doing this include the following:

- **Improve healthcare** so that the survival rate of children increases. With higher survival of children, parents sense less need to have as many children.

- **Increase education**, which will increase the options young adults have in their childbearing years.

- **Increase social security for the elderly** so that parents do not sense that they must have a lot of children to take care of them in their old age.

- **Increase career opportunities for women,** which has the effect of increasing generation time and slowing growth, and may remove some women from child-bearing altogether.

- **Extend generation length,** which decreases overall population growth. This can be accomplished through education and providing career opportunities for women, so that there is social status for accomplishments other than bearing children.

- **Decrease birth rates**, which is often accomplished through contraception. (In the case of China, the government's attempt to decrease birth rates led to a "one-child" policy that had unintended social consequences.)

- **Increased perception of wealth** will prompt parents to invest more in each child, have fewer children, and feel more secure in their old age.

Case Studies and Policies on Managing Population Growth

1. The **United States** has not directly managed its demographic transition, but it has modeled possible strategies for other countries, which include contraception, improved healthcare, increased education, lengthened generation times, social security for the elderly, and increased opportunities for women. With continued population growth, the average age of marriage continues to climb.

2. In 1979, **China** attempted to limit its burgeoning population with a "one-child policy" in urban areas. This policy met with mixed outcomes. Currently, the country is becoming more industrialized, which is prompting many strategies that lead toward a demographic transition. Current annual population growth is less than that of the United States.

3. **India** has been very deliberate about prompting a demographic transition, using strategies such as introducing family planning and birth control programs, and building a perception of wealth through industrialization, education, and technology development.

4. **Africa** is much less unified than the above countries, and has struggled with disease—particularly AIDS—becoming endemic, and a major form of environmental resistance. Individual countries have enacted various policies with some good results, such as improved healthcare, family planning, and industrialization.

The Effects of Population Growth

Shifting carrying capacity—The carrying capacity of our global environment can change if pollution reduces our society's ability to produce and transport food, survive disease, settle new places, and live close together. Technology has the potential of increasing carrying capacity by increasing crop yields and improving our ability to fight disease. But there can be an environmental cost to technology that could decrease carrying capacity. Weather disasters also have the ability to sharply decrease carrying capacity; consider the effects that Hurricane Katrina had on the Gulf region. If an area's carrying capacity decreases while population increases, the negative impact of population growth accelerates.

Hunger—While there seems to be enough food to feed the current global population, our ability to distribute that food to the neediest people is not efficient enough to prevent world hunger. Many countries have more food than they need. Some countries have enough food—but not the right collection of nutrients—and are therefore malnourished. In addition, climate change and poor management of croplands are decreasing the amount of land available for growing food.

Disease—Greater population densities, combined with the shifting distribution of disease vectors due to global warming, are increasing the rates of disease. Some scientists contend that higher background levels of pollution are contributing to greater incidences of blood cancers, neurological deficiencies, and toxic compounds in our food. Increased catastrophic events, such as hurricanes and tsunamis, in densely populated areas devastate essential water and wastewater infrastructure, which contributes to outbreaks of cholera. High population densities, combined with the ease of global travel, allow emerging diseases to spread quickly—having a greater impact—before human populations can adapt and less virulent forms of the disease can develop.

Resources—With an increase in human population, the demand for resources also increases. In the search for more essential resources—such as coal to produce electricity, or the development of croplands or watersheds—more and more land is devastated. As more countries develop, the amount of resources used by each individual increases.

Habitat destruction and the subsequent loss in biodiversity is a major threat posed by human population growth.

> **DID YOU KNOW?**
>
> Easter Island is noted for its magnificently large stone statues of giant faces. When European explorers found Easter Island in 1722, they found a poverty-stricken population and could not understand how such a weak population could produce such fabulous monoliths. Most of the remaining inhabitants were dead only a few years later. Scientists now believe that Easter Island was a lush island with a thriving culture about 2,000 years ago. However, the population of the island exceeded its carrying capacity, based on the resources available on the island. The culture vanished as a result of indulgence in a lifestyle that was not sustainable.

Time for a quiz
- Review strategies in Chapter 2
- Take Quiz 3 at the REA Study Center
  (www.rea.com/studycenter)

Take Mini-Test 1
on Chapters 4–6
Go to the REA Study Center
(www.rea.com/studycenter)

Chapter 7

Land and Water Use

The last two chapters focused on basic principles in the Earth and biological sciences. Those topics are often found in greater detail within other courses that some students may have taken. (Don't worry, however, if your AP Environmental Science experience is your first exposure to these topics!) In this chapter, you will review topics that are unique to the AP Environmental Science syllabus. You will apply the consequences of human action to those fundamental sciences you covered in the previous chapters. This is where your AP Environmental Science course diverges from other science courses, and where your ability to synthesize ideas from multiple chapters will really pay off. In this chapter, you will review concepts of sustainability as they apply to many different human uses of land and water.

Relevant laws and treaties, as well as sustainable land-use strategies, are distributed with the particular topic covered, rather than collected in one section. Be sure to learn at least one relevant law with each general topic, as well as sustainable practices for all uses.

The College Board syllabus indicates that the following topics from this chapter may show up on the AP Environmental Science exam. These topics compose about 10% to 15% of the total test.

  A. **Agriculture**
   1. Feeding a growing population (Human nutritional requirements, types of agriculture, Green Revolution, genetic engineering and crop production, deforestation, irrigation, sustainable agriculture)
   2. Controlling pests (Types of pesticides, costs and benefits of pesticide use, integrated pest management, relevant laws)

  B. **Forestry**
   (Tree plantations, old-growth forests, forest fires, forest management, national forests)

C. **Rangelands**
(Overgrazing, deforestation, desertification, rangeland management, federal rangelands)

D. **Other Land Use**
1. Urban land development (Planned development, suburban sprawl, urbanization)
2. Transportation infrastructure (Federal highway system, canals and channels, roadless areas, ecosystem impacts)
3. Public and federal lands (Management, wilderness areas, national parks, wildlife refuges, forests, wetlands)
4. Land conservation options (Preservation, remediation, mitigation, restoration)
5. Sustainable land-use strategies

E. **Mining**
(Mineral formation, extraction, global reserves, relevant laws and treaties)

F. **Fishing**
(Fishing techniques, overfishing, aquaculture, relevant laws and treaties)

G. **Global Economics**
(Globalization, World Bank, Tragedy of the Commons, relevant laws and treaties)

Agriculture

Farming is much more complicated than you may think! Human agriculture is an outcome of the Neolithic Revolution, and has the challenge of providing ongoing human nutritional needs for a growing population in a manner that minimizes disease. To best understand the challenges of agriculture, you should understand essentials of human nutrition, basic agricultural methods, and risks of disease and environmental health that makes agriculture less sustainable.

Feeding a growing population is a global challenge because humans need enough calories, amino acids, and vitamins/minerals to survive.

1. **Basic human nutritional requirements**
   - **Essential calories**—Humans need 2,000 to 2,500 calories each day. To get less is considered *undernourishment*.

- **Essential amino acids**—Humans need 22 different amino acids in order to build all the proteins that are needed by the body. Proteins are used for critical functions in every cell. Of these 22 amino acids, the body can synthesize all but eight. These eight are considered essential amino acids that must be in each person's diet. If a person's diet is missing any one or more of the essential amino acids, then the building blocks necessary for healthy living are not present. Meats, cheeses, and eggs contain all eight essential amino acids. They can also be obtained from combinations of plants, such as legumes (beans) and a grain (wheat).

- **Essential vitamins and minerals**—Essential vitamins include vitamins A, B, C, D, E, K, and folic acid. The essential minerals include calcium, phosphorus, magnesium, sodium, potassium, iron, zinc, and fluorine. The essential trace elements include copper, chromium, manganese, molybdenum, selenium, and iodine. The essential vitamins and minerals combine with other molecules—particularly proteins—to catalyze chemical reactions in our bodies, operate nerves and muscles, and create a stable physiologic environment with our bodies.

> **DID YOU KNOW?**
> Agriculture in different cultures has focused primarily on supplying the eight essential amino acids, and secondarily on providing other essential vitamins and minerals.

2. **Undernourished versus malnourished.** While the inability for humans to obtain enough calories is considered undernourishment, humans who do not acquire adequate vitamins and nutrients are considered malnourished and will eventually develop a disease associated with the nutrient that they are missing.

    - **Kwashiorkor** is easily recognized in children with swollen abdomens and reddish orange hair, and is the result of a lack of protein in the diet.

    - **Marasmus** results from the lack of both protein and total caloric intake, and causes skeletal thinness, wrinkled skin, and decreased immunity to disease.

    - **Anemia** results from a lack of iron, which often comes as a secondary result of a lack of animal protein. Anemia prohibits oxygen from traveling to tissues and results in low energy and fatigue.

    - **Ariboflaninosis** results from a deficiency of vitamin B_2 and is one of the most common nutritional disorders in the United States. Symptoms include skin problems, sore lips and mouth, and anemia.

    - **Goiter** and **hyperthyroidism** result from an iodine deficiency.

- **Rickets** results from the body not having enough calcium in the bones, which is often brought on by a deficiency in vitamin D.
- **Pellagra** results from a deficiency of niacin, which can produce skin problems, learning difficulties, or even death.
- **Scurvy** results from vitamin C deficiency. Symptoms include loose teeth, black-and-blue spots on the skin, and swollen gums.
- **Beriberi** results from thiamine (vitamin B_1) deficiency. Symptoms include a loss of appetite and cramps.

3. **Toxins in food** have become an increasingly serious environmental threat as humans seek to raise food in a cost-effective manner. The following are examples of threats that can introduce a toxic component to food:
   - **Antibiotics** are used in many agricultural settings, particularly in raising animals used as a meat source. Antibiotics allow growers to raise animals in densely populated factories. However, the antibiotics not only remain in the meat, but work their way into the water cycle and the rest of the food chain, resulting in more bacteria and diseases that are resistant to antibiotics.
   - **Pesticides** in food may help increase productivity, but they present a risk to non-target species, such as humans. Humans can experience toxic effects from either acute (short-term) or chronic exposure. Acute exposure at high levels may cause nerve problems, bleeding, or death. Chronic exposure is suspected to cause decreased immunity, cancer, birth defects, and other diseases. Pesticide use is examined in more detail later in this chapter.
   - **Biomagnification** occurs when toxins accumulate in a producer or primary consumer and are concentrated as they are ingested and further retained in subsequent trophic levels.
   - **Bioaccumulation** occurs within a trophic level when fat-soluble or radioactive toxins remain in body tissue and build up over time.

4. **Famines** are massive, acute incidences of undernourishment that are usually catalyzed by political or economic upheaval (as in war) or environmental devastation (such as a storm or a tsunami). In either case, while the acute nature of the famine may pass quickly, the ability of a region to return to full productivity is severely weakened.

Relevant Laws

The **Federal Food, Drug, and Cosmetic Act (FFDCA, 1938)** sets tolerance limits for toxic residues in food, drugs, and cosmetics. The Delaney Amendment requires absolute lack of toxic hazard for food and drugs and defines "acceptable risk" as one case of cancer in 1 million exposures.

> **TEST TIP**
>
> While the concepts of biomagnification and bioaccumulation show up here in the context of toxins in food, they are very important environmental concepts that can be applied to other topics, such as toxic waste. Be sure to understand these topics well.

Agricultural Methods

1. **Irrigation methods** continue to develop in ways that match the delivery system with the amount of water available. Sprinkler techniques spray water over crops using movable pipes, fed by deep wells. These techniques use a lot of water and quickly deplete groundwater. Drip irrigation applies water directly to the base of the plant or tree through low-flow hoses. Less water is wasted, but the systems are less cost effective for very large plots. Flood irrigation covers an entire field with water for a brief time and saturates the field. It is cost effective for some crops, such as hay and rice, but it may contribute to erosion or waterlogging if done poorly.

2. **Soil conservation** became much more prevalent after the Dust Bowl era of the 1930s. Soil conservation practices include using low- or no-till farming practices (leaving plant material on the ground after harvest), planting trees as windbreaks, planting rows that follow a hill's contour, and monitoring soil nutrients to make sure they are not becoming depleted. The following are methods of soil conservation.

   - No-till or low-till farming—leaves plant material lying on the ground after harvest to protect the soil from wind and helps retain moisture.
   - Windbreak trees—planted to shield the soil from prevailing winds.
   - Contour farming or terracing—reduces water erosion by keeping water from flowing downhill.

- Adding soil nutrients—maximizes the adhesive quality of the soil and helps contain erosion.

3. **Genetically modified organisms (GMOs)** are organisms that have been genetically engineered to carry the gene of another organism. Scientists began developing GMOs to produce strains that are more resistant to pests, pesticides, and damage from shipping, and that could produce higher yields. Today, transgenic crops represent the majority of food grown and sold in the United States. Some consumer advocates are concerned because it is difficult to predict the health and environmental drawbacks of GMOs. There are many unanswered questions about whether or not GMOs have become virulent exotic strains, or carry proteins that may induce new allergies to previously allowed crops.

> **DID YOU KNOW?**
> The Ogallala Aquifer is the largest aquifer in North America. It lies underneath the central plains states of Oklahoma, Kansas, Colorado, Nebraska, South Dakota, New Mexico, and parts of Texas. In some places, it is over 1,200 feet thick, and provides the water for America's Breadbasket. Farmers and municipalities began to pump water from the Ogallala about 100 years ago—and now this aquifer, that took thousands of years to build up, is being depleted. Careful attention is now being paid to the rate at which the aquifer is being recharged and to the water tables in these states. We now know that we have to be careful stewards of this common resource! (See the section on Tragedy of the Commons at the end of this chapter.)

4. The **Green Revolution** began in the 1950s and developed high-yielding strains of rice that were mass-planted over large areas. Yields increased sharply for a time, but then other pest resurgence and decreased resistance to disease were discovered to be part of the cost of large-scale monoculture. The Green Revolution also brought higher dependence on agricultural chemicals, such as pesticides and fertilizer, and more expensive seeds.

5. **Sustainable methods** of agriculture include a wide range of practices that incorporate a balance of nutrients, using less energy and water, and reducing the amount of fertilizer and pesticides. Some sustainable methods include:

    - Low- or no-till farming and contour farming to decrease wind and water erosion.

    - Crop rotation and simultaneous polyculture to keep a balanced level of nutrients in the soil.

- Use of natural fertilizers — preferably organically composted organic waste — rather than chemical fertilizers.
- Minimizing pesticide use by using integrated pest management, or by switching the crops regularly to minimize pest momentum.
- Minimizing the use of fossil fuels by selling to local markets, using local materials, limiting the number of times a farmer uses machinery to disturb the soil, and decreasing the use of petrochemicals.
- Minimizing the use of water, particularly if the water comes from groundwater or from a distant water source — both of which have extreme long-term effects on the environment.

> **DID YOU KNOW?**
> The use of GMOs for foods routinely sold in grocery stores has increased dramatically in recent years. In the United States, it is now more unusual to eat food that does not contain GMOs than it is to eat food that contains them.

The Environmental Impact of Agriculture

As much as we need agriculture to support us, using unwise practices have the potential to create long-term environmental damage.

Deforestation occurs when forests are removed to create space for agriculture. Most agricultural regions are natural grassland biomes. However, in some places and times in human history, farmers have tried to expand agricultural lands into cleared forested areas. Forests tend to have less fertile soil, and the removal of trees increases erosion, decreases transpiration, and ultimately shifts the climate in the region. Deforestation of rainforests in South America have actually led to reduced rainfall and hopelessly eroded areas. (See the section called Forests later in this chapter.)

Desertification occurs when land has been overfarmed and it is sometimes accelerated by climate shifts. The nutrients and organic materials in the land become depleted and sandy soil is all that remains. When the soil is sandy, it is more susceptible to erosion, has less water evaporation occurring above it, and heats up more quickly. If this effect is multiplied over a large area, the land becomes more desert-like.

Erosion occurs when soil is moved from its point of origin. There are three types of erosion: rill erosion, where water cuts small rivulets in the soil; gully erosion, where rill erosion escalates to form a large channel; and sheet erosion, where water removes a horizontal layer of soil. Erosion is caused by wind, water, chemical reactions (including salt deposition and *salinization*), physical changes (such as compaction from overgrazing), and waterlogging from excessive irrigation or poor drainage.

Soil nutrient depletion occurs when crops continually pull the same nutrients out of soil, and the nutrients are not replenished. For example, farmers learned to rotate crops, or plant some crops next to others, so that a crop that depletes nitrogen from the soil is followed by a crop that "fixes" nitrogen in the soil. (See the discussion about the nitrogen cycle in Chapter 5.)

Relevant Laws

The **Soil Conservation Act of 1935** was passed in the wake of the Dust Bowl to prevent a repeat of the soil depletion that occurred as a result of poor farming practices.

Pest Control

Pesticide use becomes a land management issue because it greatly influences the type of agriculture that is practiced.

Types of Pesticides

Inorganic pesticides, such as arsenic, copper, mercury, and lead, are highly toxic and persistent.

Chlorinated hydrocarbons, such as DDT, aldrin, lindane, and toxaphene, block nerve membrane ion transport and the nerve signal; they tend to bioaccumulate and biomagnify.

Organophosphates, such as parathion and malathion, are strong neurotoxins, but they are not persistent.

Carbamates, such as carbofuran and aldicarb, behave much like organophosphates.

Botanical pesticides, such as pyrethrum, which is extracted from chrysanthemum flowers.

Costs and Benefits of Pesticide Use

Costs include toxicity to nontarget species—especially through bioaccumulation and biomagnification, pest resistance and resurgence, and increased financial cost. External costs may include health costs of farm workers and environmental costs outside the immediate area.

Benefits include higher yields, at least at first.

Pest Management

Integrated pest management (IPM) systems involve a combination of pest control strategies—chemical and nonchemical—that are unique to the crop and location. Non-chemical strategies include the use of natural predators, the use of sex pheromones to attract bugs to a place away from the crops, the introduction of sterile breeding partners, and crop rotation. Chemical products can also be part of an IPM, but they are used with greater deliberation and specific targeting so that a minimum is used.

Relevant Laws

The **Federal Insecticide, Fungicide, and Rodenticide Act (FIFRA, 1947)** regulates the manufacture and use of pesticides.

Rangelands

Rangelands are large grazing areas that include woodlands, grasslands, wetlands, deserts, savannas, chaparrals, and tundra. Rangelands grow native flora, rather than being cultivated by humans with seeding, irrigation, and the use of chemicals. Rangelands are managed primarily to maximize grazing potential in a natural setting.

Rangeland management focuses on providing leases for private ranchers to graze cattle or sheep on public lands in a manner that does not destroy the range. Federally-owned rangeland is overseen by the Bureau of Land Management (BLM), which is within the Department of the Interior. The BLM manages about 260 million acres of open rangeland in the United States.

Overgrazing is a constant issue because ranchers are motivated to graze the largest number of animals within their lease agreements. If done excessively, particularly if there is also a change in climate, overgrazing can lead to deforestation and desertification.

Deforestation occurs in some countries as ranchers purchase forests, then cut them down to provide room for cattle grazing. In this quest for short-term economic gain, ranchers are spending the "natural capital" of the region that is difficult to repair. Deforestation reduces the available moisture in an area and increases erosion of nutrients from the soil. While grasses can be supported for a time, the soil is rarely fertile enough to sustain a grassland indefinitely, and the region progresses into desertification.

Desertification is the process of converting a nondesert ecosystem—such as grasslands or a forest—into a desert. In some developing countries with growing populations and limited resources, overgrazing rangeland and overuse of available groundwater accelerates the devolution of grasslands to desert areas. Cattle graze on and eat the grass, and future grass has a difficult time growing with less moisture and nutrients. Desertification has accelerated in the Sahel region in Africa (which has contributed to famine in that area) and also in South America.

Fishing

Although not commonly thought of as agriculture, fishing has increasingly taken on such a carefully managed role and it is so critical for capturing nutrition that it fits best in this category for review purposes.

Fishing Techniques

Purse seining catches fish by towing a large net between two boats, then closing the net at the bottom. Like other methods, the by-catch is high. However, if used for running salmon, the seining operation rarely catches fish other than salmon.

Gill netting involves nets that are typically set in a straight line, tangling fish in the net by their gills. Gill netting is more selective than purse seining, long lining, and trawling because the size of the mesh is proportional to the size of the fish caught. There is, however, still considerable by-catch associated with gill netting.

Hooks attached to a long single line, or used with a handheld pole, often capture species not intended to be caught, which may include sea turtles and birds.

Trawling drags nets through the water. Trawling is sometimes done by dragging a net between two boats, or a net is dragged along the bottom by a single boat. Like seining, trawling is notoriously unselective in capturing fish, catching fish of illegal size and species, along with marketable fish.

Traps are used to catch lobsters and crabs, and other benthic invertebrates.

Overfishing, catching so many fish that they cannot sustain their population, is a major concern of the commercial fishing industry. Some estimates suggest that more than 70% of fish species are overexploited or depleted.

Aquaculture is one possible answer to correcting depleted fishing resources. Aquaculture may occur in large oceangoing fish farms, in buildings, or in shore- or land-based ponds and lakes. Increasingly, salmon are farmed in large, open-ocean aquaculture "farms" that contain the salmon, which are fed artificially. Some feel that these salmon are not as healthy for consumption, and some unintended products—such as PCBs and antibiotics—bioaccumulate in the farmed salmon. Another concern is that the genetically-modified salmon may escape and "infect" wild populations, like other exotic species invasions. Others feel that open-ocean farming is one of the best ways to allow natural fisheries to thrive.

Sustainable fishing takes into account the overall impact of fishing on the marine environment, fish population dynamics, and methods that limit destroying nontarget populations. There is a new push to provide a "sustainable fisheries certification" to fish that are harvested in a sustainable manner, so that consumers who care about sustainability can support those fisheries.

Relevant Laws

The **Fish and Wildlife Act (1985)** establishes that fish and wildlife resources be given "equal consideration" in balance with other uses of water resources.

The **Magnuson-Stevenson Fisheries Management and Conservation Act (1996)** governs all fishing in U.S. waters up to 200 miles from the nation's coast. This act promotes good management of commercial and recreational fishing resources.

Forests

Forests and carbon sequestration. Forests undergo a very high level of photosynthesis, which uses carbon dioxide and water, gives off oxygen, and stores carbon in the form of carbohydrates and structural molecules. Consequently, forests are one mechanism by which carbon is removed from the atmosphere and stored in a solid form. This type of storage by forests is an example of carbon sequestration (see Chapter 5).

Tree plantations have been developed on many continents to increase lumber yields. However, tree plantations limit biodiversity, and often involve some of the same drawbacks as other types of large-scale monoculture: pest and disease resurgence, use of chemicals, loss of biodiversity, and soil degradation.

Old-growth forests are forests that have not been disturbed by humans and typically represent a mature community. Old-growth forests contain a rich diversity of species and trees, which often provide a multilayered canopy. They also have a high biodiversity of flora and fauna, which would not necessarily thrive in other forests. Old-growth forests, such as the redwood forests in northern California, were first used as a source of lumber. The wood harvested from an old-growth forest has a higher density than the wood found in forests specifically cultivated for timber harvesting.

Deforestation has occurred in many areas as a result of the following:

Fuel gathering has been a major reason for deforestation in developing countries and has had a large impact on whether or not entire cultures survive through history.

Urbanization and farming has pushed back forest boundaries to make way for more people and commercial activity.

Clear-cutting is a logging practice that cuts down every tree in an area, regardless of species or size. The soil is exposed to erosion and the forest habitat is essentially destroyed. Large slag piles of waste wood are pushed up with bulldozers. While replanting helps and is a noble effort, it is difficult to replicate the same density of forests using artificial methods. *Selective cutting* is less cost-effective, but it involves a less disruptive harvesting of mature trees, which results in better growth, continued habitat for wildlife, and less erosion.

Swidden agriculture refers to a practice used by indigenous people in tropical rainforests. The farmer clears a small plot of land by cutting, and often burning, trees. The ashes from the fire provide essential nutrients, which do not erode away because neighboring trees hold in the soil. A mixture of crops, such as bananas and sweet potatoes, are planted together. Swidden agriculture is sustainable as long as the land is not used too quickly, and the farmer allows each plot to undergo secondary succession, and then moves on to another plot of land. However, if the density of farmers is too high, or if outside companies open up large tracts of rainforest, then the forest does not recover and deforestation occurs.

> **DID YOU KNOW?**
>
> Deforestation seems to have been responsible for the decline of many cultures, including the Anasazi in ancient North America and the Mayan civilization in Central America.

Sustainable forestry is accomplished when forestry operations take sustainable practices into account. As a result, lumber that comes from sustainable forestry practices is certified so that consumers know that they are purchasing wood grown and harvested in a sustainable manner. Sustainable forestry practices take into account growing and harvesting trees while maintaining the surrounding ecosystem. This practice keeps the forest healthy and its biodiversity high, protects the forest for the future, and uses practices that decrease erosion and its effects.

Fire management has some benefits compared to putting out all fires. For many years, the policy of the Forest Service was to put out all fires. However, this policy allowed the undergrowth of the tree canopy to grow unhindered, which resulted in hotter and more damaging fires when they did come. Additionally, eliminating all fire from some forests, particularly the taiga forests in the northern Rockies, also eliminated the benefits that fire provide to many species. For example, a fire opens the cones of the Lodgepole Pine and facilitates reseeding. Fires open up areas and help with meadow formation, which allow grazing wildlife to move to new areas. Now the Park and Forest Services have a "let burn" policy that uses firefighting resources only when lives or property are threatened. This new policy attempts to allow smaller fires to remove the undergrowth and revitalize the forest community.

Wetlands

Wetlands are mixed terrestrial/aquatic areas that are under standing water or saturated for a significant portion of the year. Wetlands include fens, bogs, marshes, riparian zones, and swamps. While formerly thought of as a nuisance, wetlands are now considered to be a significant biological asset because they support enormous biodiversity, work as "nature's kidney" by cleansing toxins from the water that passes through, and provide recreation for those who wish to view wildlife. As you review the importance of wetland biomes, keep the following in mind:

Water purification is one of the largest benefits of wetlands. Water slows down as it passes through the wetlands and pollution is metabolized by microbes and plants.

Flood and erosion control represent one of the major utilitarian advantages of wetlands. When rivers have a nearby wetland to receive excess runoff, there is less chance of flooding.

Habitats represented by wetlands support rich biodiversity because of their nutrient rich soil and protective spaces. Decomposing microbes become food for zooplankton, which are eaten by insect larvae, which in turn become food for birds and small amphibious reptiles. The abundance of flora and fauna help wetlands support migrating animals, such as birds.

Nutrient pollution occurs when wetlands are not able to use up nutrients like nitrates and phosphates. These nutrients enhance eutrophication, which contributes to decomposition and, later, build up of biomatter, which accelerates succession into meadows and grasslands.

Habitat destruction due to development has diminished wetlands significantly, particularly in states like Florida, where the climate is pleasant. Numerous wetlands once dominated the state.

Conservation of wetlands is promoted by numerous acts of legislation. Wetland conservation has taken on further economic significance because developers will purchase wetlands in one location and use as a trade, or mitigation, of wetlands they would like to develop in another location.

> **DID YOU KNOW?**
> Steady use of Florida groundwater and the ever-increasing diversion of water have reduced the high water table in and around the famed Everglades National Park. Steady development has converted many thousands of acres of wetland into residential and farming areas. Not only does this reduce the flow of water into the Everglades, but runoff from these developed areas sends nutrient pollution and toxic wastes into the Everglades—endangering the species protected by the park.

Mining

Mining and drilling operations extract mineral deposits, solid metals (such as gold, silver), ores (rocks that contain minerals), and hydrocarbons. Minerals, solid metal, and ores have nonbiogenic sources, whereas hydrocarbons (oil, gas, coal) were originally living things that decayed into mixtures of hydrocarbons and—in the case of coal—just carbon. (Hydrocarbon formation is described in detail in the next chapter.)

Mineral formation occurs in one of three ways: when hot molten rocks cool, when two soluble ions combine to form an insoluble mineral, and when water dissolves and leaves otherwise soluble minerals. Minerals come from ionic compounds with homogenous chemical properties for each mineral. By contrast, rocks are heterogeneous mixtures of minerals.

Extraction is the point during the mining process that actually takes material from the ground. The entire mining process includes planning and due diligence, extraction, processing, and reclamation. The following three methods are the most common forms of extraction:

1. **Surface mining**—To dig an open mine, the overburden is first removed and stored. Then the mineral seam is mined. In some cases whole mountains are displaced, or open pits several kilometers across are created.

2. **Subsurface mining**—Underground mines are used to extract deeper mineral deposits. This method presents many more risks to miners—gases, lack of oxygen, fires, explosions, and tunnel collapse. Also, mine tailings (the refuse that remains after processing) tend to be removed and discarded on the surface. Water percolating through the tailings creates a toxic, acidic runoff that can pollute streams and waterways.

3. **Wells**—For gaseous and liquid deposits, a well may be the best way to extract the resource. *Fracking* is a new method used to obtain more product from a well. Fracking involves mixing water with chemicals and pumping the solution into the ground in order to force out more product. The process is coming under greater scrutiny because some say it is resulting in polluted groundwater, hydrocarbons in well water, and geological disturbances.

> **DID YOU KNOW?**
> Fracking can infuse so much natural gas into groundwater, that tap water in some areas where fracking was used can actually ignite and burn!

Environmental consequences of mining can include deformation of the land, erosion from physical or chemical means, pollution of ground water from internal sources or exposed mine tailings, toxic particulate matter, and lost use of land.

Mineral rights represent ownership of the minerals underneath the surface. When mining operations seek land to mine, they must have permission—and pay royalties to—the person or entity that owns the mineral rights. Most of the time, the person who owns the land does *not* own the mineral rights. Therefore, mining operations sometimes involve serious imposition on the landowner because the landowner has no rights to keep a mining operation from mining the area below the surface.

Global Reserves

The following table shows the important mineral reserves and major producing countries in the world.

Table 7-1. Worldwide Mineral Reserves

Mineral	Uses	World Reserves (Metric Tons)	Countries
Bauxite	Ore of aluminum	21,559,000	Australia, Jamaica, Brazil
Chromium	Alloys, electroplating	418,900	India, South Africa, Turkey
Copper	Alloys, electric wires	321,000	Chile, United States, Canada
Gold	Jewelry	42	South Africa, United States, Australia
Iron ore	Iron and steel	64,648,000	Brazil, Australia, China, Canada, Venezuela
Lead	Solder, pipes	70,440	United States, Mexico, Canada
Manganese	Iron and steel	812,800	South Africa, Gabon, Australia, France
Nickel	Stainless steel	48,660	Canada, Norway, Dominican Republic
Silver	Jewelry	780	Mexico, United States, Peru, Canada
Tin	Tin cans, alloys	5,930	China, Brazil, Indonesia
Zinc	Iron and steel	143,910	Canada, Australia, China, Peru, Mexico, Spain

Source: economywatch.com

Relevant Laws

Surface Mining Control and Reclamation Act (SMCRA, 1977) requires mining companies to put funds in escrow to ensure land reclamation after the mining operations have ceased. The act puts states in charge of mining reclamation.

Urban Land Development

Urbanization Trends

In 1800, 6% of the U.S. population lived in an urban setting. In 1990, 75% lived in an urban environment. Not only is our population growing, but an increasing percentage live in a city.

Reasons for immigration to cities vary. Factors that pull people toward cities include improved sanitation, improved access to cultural and consumer products, improved medical care, higher salaries and more employment opportunities, more recreation, and more luxurious housing.

There are also factors that push people away from rural areas, toward cities. Examples of this include rural areas having less social diversity, less entertainment, less economic vitality, and less medical care. Additionally, much of the land may be divided so much that rural plots may not be large enough to support a family.

Urban Environmental Problems

1. **Traffic** is a struggle for many cities. Some have developed public transportation systems, but some have experienced such rapid growth that planning and funds have not kept pace, and city dwellers spend more of their day inside a car than they might in a more rural setting.

2. **Air pollution** is certainly an issue in most cities because of the industrial practices nearby. Traffic, industry, and energy production (from coal, for example) all contribute to increased air pollution in cities. In developing countries, lenient standards, corrupt governments, and lack of funds make urban air quality even worse.

3. The **heat island effect** creates a bubble of heat surrounding the city because of the heat given off by buildings, cars, and energy production. These heat islands hold in pollution and increase air-conditioning demands in warm weather—furthering the economic and environmental impact of the city.

> **DID YOU KNOW?**
> Between our desire to work in cool buildings and the heat island effect, most American cities use more electricity during the summer than they do during the winter, primarily to operate air conditioning.

4. **Water pollution** is significant because it is difficult to treat all the human waste that is generated in a densely populated city. Local rivers carry the burden of nutrient pollution from treatment facilities, which are sometimes overwhelmed at times of high runoff during storms. When multiple cities are developed along the same river, the waste created by an upstream city often becomes the downstream city's drinking water.

5. **Crime** is notoriously higher in cities, where people can become increasingly desperate for comfort, or just survival. Robbery or prostitution to gain money for food or drugs, and violent actions to vent frustrations or injustices, are made easier because of the anonymity of the city. Fewer extended families to provide the support for those on the margin increase the incidents of substance abuse and related crimes, and also the ambient level of mental illness.

Suburban sprawl developed in the 1950s through the 1980s as people attempted to raise families in large homes with surrounding land, but still work in the cities. Suburban sprawl increased the cost of electrical, water, and sewage infrastructure per citizen, and required people to use cars for all activities.

Sustainable Urbanization

If managed well, urban settings can actually be more environmentally sustainable because there are lower levels of resource use, per capita, than if people live in less dense settings. In urban areas, people are more likely to use public transportation, use less water (thus lessening the waste infrastructure needed per person), and require less land. Some sustainable practices that can be found in newly developed urban areas include:

1. **Gardens on rooftops** and **parks**, some of which provide locally grown vegetables.

2. **Transportation** that allows people to get around without a car. Some cities are using light rail and trolleys to connect people to shopping and work. Less fuel is used, and less pollution is produced.

3. **LEED-certified buildings** that minimize the impact on the environment and use less energy. (See the section called LEED Certification in Chapter 9.)

4. **Shops and open space** that allow people to meet their needs, and experience pride and serenity, without traveling too far.

5. **Multi-use design and clustering** that places attractive living spaces close to, or on top of, retail, office, and even light industrial spaces.

6. **Small business spaces** nearby that provide diversity of experience, as well as personal and professional growth.

7. **Social equity** in housing that mixes economic groups can help lessen crime.

> **DID YOU KNOW?**
> The U.S. Green Building Council has instituted a program for buildings called Leadership in Energy and Environmental Design (LEED). Buildings that have LEED certification have less impact on the environment and use less energy than traditional buildings. Many urban developments, such as the new downtown development in Portland, Oregon, are composed entirely of LEED-certified buildings, which draw to them people who are interested in living in structures that are environmentally friendly.

Transportation Infrastructure

The Federal Highway System

1. **The interstate highway system** was originally funded by the Federal-Aid Highway Act (1956) to facilitate the movement of military resources and aid in evacuation of major metropolitan areas. Today, it covers over 47,000 miles and is critical for commercial truck transport and for individual travel for business and recreation.

2. **Change in demographics** has occurred as people do business and move along major interstate corridors.

3. **Environmental costs** of the highway system include land degradation from the millions of tons of gravel used for the road bed, the use of millions of tons of hydrocarbon-based oil and asphalt, and the interrupted migration and genetic isolation of wildlife.

Channels and Canals

1. **Channels** are the deeper section of a waterway that is more frequently used for navigation. In some rivers, channels are often marked so that boats and ships do not get stranded in shallow water. The Ohio River has been cut and dredged in such a way that it is navigable by coal barges and tugs.

2. **Canals** are waterways that have been cut between bodies of water, such as the Erie Canal or the Suez Canal.

3. **Environmental costs** include the migration of invasive species into major bodies of water, such as the Great Lakes, and *channelization,* which leads to bank erosion, loss of wetlands, and increased flooding. Canals used to transport water long distances—such as the Aral Sea in Russia and the California Water Project—have good initial intentions of increasing farm fertility, but they eventually contribute to ecosystem degradation, including soil salinization.

Roadless areas take up about 58 million acres in national forest areas (see the section called Public and Federal Lands below). While roads are very useful for commercial and recreational reasons, maintaining these areas without roads is important for species and ecosystem protection.

Sustainable transportation maximizes the amount of transportation for the largest number of people using the least amount of materials and energy. Transportation is responsible for about 25% of the world's use of energy and production of carbon dioxide, which is a greenhouse gas. Therefore, considering solutions that use less energy and emit less pollution has a significant impact on the environment. Some possible methods of transportation that are more sustainable than an individual driving a car include the following:

> **DID YOU KNOW?**
> The Ohio River spans several states in the American Midwest. Along the river are 19 lock and dam combinations, seven of which are also hydroelectric facilities, all built by the Army Corps of Engineers years ago. The combination of locks and dams allows coal barges to navigate up and down the river, and transport coal from mines in Pennsylvania and Kentucky to 47 coal-fired power plants along the river. In addition to navigability, the system helps control flooding. Some would argue, however, that the channelization of the river has reduced wetlands and actually increased flooding, and the power plants have increased thermal pollution, acid rain, and mercury deposition in the region.

1. **Walking and cycling** are becoming increasingly popular, and are leading some people to live closer to work so that they may have a healthier, sustainable lifestyle.

2. **Hybrid and electric cars** have greater fuel economy and therefore do not use as many fossil fuel resources as traditional cars. Hybrid cars contain batteries, which are charged when the car brakes or coasts, and supplement the mileage of the car's internal combustion engine. Electric cars are fueled at charging stations that are becoming increasingly available.

3. **Ride sharing** helps reduce emissions and increase the lifespan of cars

4. **Light rails, trolleys, and buses** are being used in more and more urban areas to help connect living, work, and recreational opportunities. These options use far less resources than an individual using a car, and also open up urban space.

Public and Federal Lands

Public lands are managed by departments that are part of the executive branch of the federal government. Each department is led by a member of the president's cabinet.

There are many other parts of the executive branch that manage environmentally sensitive policies. For example, the Office of Environmental Management within the Department of Energy; the National Oceanic and Atmospheric Administration within the Department of Commerce; the Food and Drug Administration and National Institutes of Health within the Department of Human Services; and the Occupational Safety and Health Administration within the Department of Labor. There are also independent agencies, such as the Environmental Protection Agency and the Nuclear Regulatory Commission, that deal with environmental issues.

The lands listed below are managed by either the Department of Agriculture or the Department of the Interior.

Table 7-2. Lands Managed by Federal Agencies

Type of Federal Land	Federal Agency	Purpose	Examples, Scope
Land managed by the Bureau of Land Management	Dept. of the Interior	Can be leased for logging and mining/drilling; often used for hunting, recreation	Most BLM land is in the west; composes about one-eighth of the U.S.
Wilderness areas	Forest Service in the Dept. of Agriculture; also Dept. of the Interior.	Established by Wilderness Act (1964). Protects wildlife and habitat. Minimizes human impact. Does not allow mechanized vehicles or tools.	Wilderness areas compose land area about the size of the state of California. Jedediah Smith (Tetons) Wilderness
National parks	Dept. of the Interior	Established by the National Park Act (1916). Preserves natural and historical landmarks. No mining, drilling, or logging. Follows preservation philosophy of John Muir.	Yosemite Yellowstone Grand Canyon Arches Mt. Rainier Glacier
National forests	Dept. of Agriculture	"Greatest good for greatest number of people" philosophy matches utilitarian conservation philosophy of Gifford Pinchot. Used for both economic development — including logging, mining, drilling, energy production, as well as recreation.	Often surround/cushion national parks; 130 national forests compose about 190 million acres in 40 states.
Wildlife refuges	Federal Wildlife Refuge system is part of the U.S. Fish and Wildlife Service, which is within the Dept. of the Interior	Protect specific species and habitats. Established by Teddy Roosevelt in 1901.	Over 500 refuges represent every ecosystem in the U.S. Wetlands protection is often a type of Wildlife Refuge.

TEST TIP

Be sure to study this chart carefully. AP Environmental Science students should be aware of the types of federal lands that protect or use environmental resources.

Land Conservation Options

Table 7-3. Options for Land Conservation

Option	Explanation	Legislation Example
Preservation	Preservation refers to providing an ample reserve of resources so that they might be enjoyed by others in the future. It is different from conservation in that preservation protects actual resources, while conservation simply decreases the use of resources.	The Park Service Act of 1916 sought to preserve natural features, unique populations, and historical objects "by such means as will leave them unimpaired for the enjoyment of future generations."
Remediation	Remediation refers to using chemical, biological, or physical methods to remove chemically active (hazardous or toxic) pollutants. Chemical treatment may include neutralization of acids or oxidants. Biological treatment may include bacterial digestion of oil, or using water plants to remove nitrates from wastewater. A physical example includes the vaporization of hydrocarbons in an underground oil spill.	CERCLA (1980, modified 1984) established a "superfund" that would begin cleaning up toxic waste dumpsites across the country.
Mitigation	Mitigation is a general term that refers to finding a solution to a problem. In the context of environmental degradation, mitigation usually refers to establishing another ecosystem elsewhere of comparable health and size, in exchange for damage done as a result of developing a nearby ecosystem.	The Fish and Wildlife Conservation Act of 1980 prompted states to develop plans to conserve wildlife resources through mitigating land exchanges.
Restoration	Restoration brings a damaged ecosystem back to its unspoiled, natural condition.	The Estuaries and Clean Waters Act establishes an Estuary Habitat Restoration Council that reviews restoration projects in estuaries.
Reclamation	Reclamation returns massively scarred land back to a condition that is useful and acceptable; the land does not have to be returned to its original unspoiled condition.	The Surface Mining Control and Reclamation Act (1977) requires a mining company to put into escrow enough money to pay for reclaiming mined lands after mining operations are complete.

TEST TIP

Be sure to study this chart carefully. AP Environmental Science students should be keenly aware of the difference between methods by which land is repaired and managed.

Water Resources Management

Water compartments—Water is stored in the following compartments around the globe. Notice which of these represent fresh, nonsalty water. Compare the water available in the freshwater compartments with water-use trends in the following section.

Table 7-4. Comparison of Water Storage Compartments

Water Compartment	Percent of Global Water Supply	Average Residence Time
Ocean	97.6	3,000–30,000 years
Ice and snow	2.07	1–16,000 years
Groundwater	0.28	Days–1000s of years
Soil, plant moisture	0.01	Weeks
Lakes and reservoirs	0.01	1–100 years
Atmosphere	0.001	8–10 days
Wetlands	0.003	Months to years
Rivers and streams	0.0001	10–30 days

Trends in water use show that water is used by three general segments of society: municipal/residential, industrial, and agricultural. Although water use trends vary from country to country, the following chart gives approximate worldwide water-use trends. Notice that the agricultural segment uses, by far, the most water.

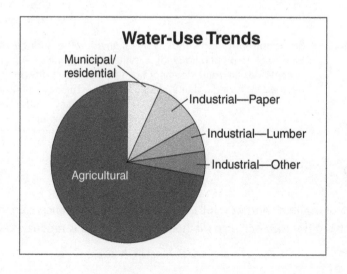

- **Industrial water use** represents a major segment of the chart on the previous page. While it is clear where water is used in one's home, it is less clear how water is used by the products we use. For example, we know we use a lot of water if we leave the sprinkler running; but we might be less sure about the amount of water that goes into the different foods we eat. Paper and pulp industries use the most water by far, followed by wood/lumber. Mining rounds out the top three industrial users of water.

- **Agricultural water use** represents the segment that uses the most water. Many farmers are switching to low-use irrigation methods. For example, instead of using large rotating sprinklers that lose most of the water to evaporation, some crops allow for the conversion to drip irrigation, which allows most of the water to go directly onto the plant and less is lost to evaporation. As the world's population grows, many regions may have to consider switching to less water-intensive crops. For example, wheat uses less water than potatoes. California and Texas, our two largest agricultural states, but also two of the driest states, may have to reduce the amount of rice and cotton they grow—two crops that require a very large amount of water.

- **Residential water use** is summarized in the chart below. As the chart suggests, most residences use water primarily to wash away waste and water lawns. Only a small portion is used for cooking and drinking.

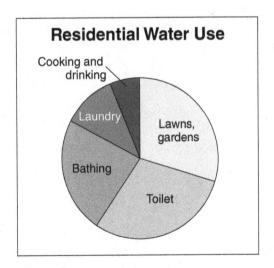

Managing Water Resources

Lock and dam projects help to support river navigation and reduce flooding. For example, the Ohio River spans several states in the American Midwest. Along the river are 19 lock and dam combinations, seven of which are also hydroelectric facilities. These were built by the Army Corps of Engineers over 50 years ago to allow large vessels to navigate up the river—in particular, coal barges to bring coal to any of the 47 coal-fired power plants along the river. The project is also designed to reduce flooding. However, some feel that the increased channelization increases flooding by reducing wetlands along the bank of the river. While this project has generated considerable electricity for the eastern portion of the United States, it has also generated considerable air pollution and thermal water pollution.

Hydroelectric dam and reservoir projects are designed to produce electricity with clean hydroelectric power, control flooding, and increase recreation opportunities.

- The **Hoover Dam** on the Colorado River was built in the 1930s. By allowing the river to back up behind it and form the immense Lake Mead, Hoover Dam is able to control flooding downstream and consistently produces a large amount of electric power. However, the dam is over 700 feet high, and an adjacent lock to allow fish and boats to pass is unreasonable. Therefore, one of the consequences of this project is that fish are not able to migrate beyond the dam.

Photo credit: iStockphoto/Thinkstock

Figure 7-1. Hoover Dam

- The **Columbia River** and the lower Snake River in Washington State contain many of the dams in a formerly complicated series of massive hydroelectric projects. The dams are being breached in order to allow salmon to migrate upstream. Many have contended that the increase in recreation and tourism is more economically significant than the power production of those dams.

- The **Three-Gorges Dam** on the Yangtze River in China is the largest dam project in the world. This gigantic project spans a mile-wide canyon and rises 575 feet above the river. Its reservoir will extend 350 miles upstream, and it will produce 18,000 megawatts of electricity with 26 turbines. However, the ecological and social costs have been very high. Over 1 million people have been displaced by the project, causing considerable political unrest. The government has seen this as a reasonable tradeoff, given the need for the country to use electricity to develop.

Water Diversion Projects

In ancient Rome, the aqueducts that brought fresh water from the north into the city were a hallmark of Roman ingenuity and convenience. Moving water to areas of heavy use is still a large challenge and poses severe environmental challenges.

- The **Aral Sea** in the former Soviet Union was used as a source of water and fishing. A large project to channel the water from this freshwater inland sea to cotton fields in the south helped the Soviets meet their cotton quotas, for a while. However, the Aral Sea lost a critical amount of water. Eventually, the nearby climate warmed and most of the sea dried up, devastating livelihoods based on fishing, and leaving salty, unusable water. The fine particles from the evaporated sea were blown into the air and caused an increase in lung diseases for residents of the region. Salinization soon infected the land that was irrigated for cotton and the land became unusable. In an effort to increase cotton production in the short term, the combinations of ecosystems in the entire region have changed and some are toxic.

- The **California Water Project** used aqueducts to bring water from the Sacramento River near the San Francisco Bay, south 350 miles through hot, arid terrain and over a mountain range so that it could be used by communities in the Los Angeles region. A second portion of the project diverted runoff from the eastern slope of the Sierras with a similar aqueduct. While Los Angeles has developed more because of the available water, the mountain runoff was not able to enter its ancestral target, Mono Lake. As a result, Mono Lake has become an American version of the Aral Sea. Pillars of salt speckle the shore and a much smaller, brackish lake remains. Aside from the political conflicts this water project has ignited, there are several health effects of the airborne salt from Mono Lake.

Channelization

Channelization is a method used to build a river's ability to accommodate the navigation of large vessels, which allows greater economic development in a region. Channelization allows more water to flow in the middle of the river and enables vessels with larger draft, or depth, to use the river. To accomplish this, the side of the river is built up and the surrounding wetlands are diminished. However, channelization increases the flow of the river, creating greater force on the riverbank and increasing erosion—by far the largest source of water pollution. Channelization also increases the speed at which the river reaches flood stage during heavy rains, and accelerates downstream flooding. The lack of wetlands adjacent to the river has diminished the cleansing and nursery qualities of low-flowing wetlands. This results in dirtier water, carrying more toxins, with less wildlife. While channelization has been an economically beneficial management tool, it is a method that results in severe environmental costs.

Water Conservation Methods

Watershed management helps some communities increase water quality and availability. For example, Clayton County, Georgia, developed a technique to supply more people with its limited water supply. The county implemented an innovative plan to treat wastewater by sprinkling it over the ground in a pine forest. As the water passed through the soil and into the groundwater, not only did it provide nutrients for the pine forest, but it was also cleansed by microbes in the soil. Eventually, it flowed into the town's reservoir, where it would enter a water treatment plant and be used again. Such watershed management has helped communities use the local geography to clean and recycle water.

Low-flow utilities increase the amount of available water because less water is used. Utilities encourage less water use by giving credits for low-flow appliances. For example, a low-flow toilet uses less than half the water of a traditional toilet. Some homes and public areas have installed composting toilets that use no water at all. As long as the waste is highly oxygenated, it does not smell. Look back to the pie chart for residential water use. What actions could reduce residential water use by 60%?

- **Decreasing spoilage** focuses on reducing siltation, which spoils more freshwater than any other type of pollution. Siltation can be reduced by limiting the runoff from construction sites, or other places where the land has been opened up. Putting recharge zones or detention basins in and around parking lots reduces the speed and forcefulness of runoff after rains, and less silt enters local creeks.

- **Reducing water-intensive projects** saves freshwater for other uses. Some products (e.g., printing newspapers) require an exceptional amount of water to produce. Producing less increases our freshwater supply.

Relevant Laws

Clean Water Act (1972) sets the goal of creating "fishable, swimmable" waterways by setting water quality standards.

Water Quality Act (1965) established the first water purity standards and created the Water Pollution Control Administration under the Department of Health, Education, and Welfare.

Safe Drinking Water Act (1974) sets standards for municipal water systems that treat water and protect groundwater resources.

Ocean Dumping Ban Act (1988) prohibits ocean dumping of sewage sludge and industrial waste.

Global Economics

There are economic and political components to managing the environment that are important to consider. To best understand these and put them in a useful context, let's first examine a fundamental concept called the **Tragedy of the Commons**.

"The Tragedy of the Commons" is the title of an essay written by Garrett Hardin in 1968. In this seminal essay, Hardin cites those aspects of human nature that caused the overgrazing of the "commons" in early New England towns, when local herders would add just one more head of cattle to feed on the communal grazing area. With little effort on the herder's part, he could gain the benefit of having an additional animal. However, the cumulative effect of all herders "just adding one more" to use the common grazing area is that the land becomes overgrazed and unable to support any animals. Tragedy results by spreading out among many the responsibility for the common area, or, as Hardin writes, "Therein is the tragedy. Each man is locked into a system that compels him to increase his herd without limits, in a world that is limited." Exercising the freedom of choice by many individuals over a common resource leads to the lack of choice for all.

The same nature in humans is also at play today as we dip from "the commons" to use forests, groundwater, oceans, air, and other resources available to all. While the economics of the system favor the person who grabs the resources first, disaster results when everyone attempts to build wealth. Hardin articulated this aspect of human nature that has been at the center of environmental degradation in many cultures over many centuries. We continue to struggle with the economics, rights, and responsibilities of a limited resource that is available to all.

> **TEST TIP**
>
> The "Tragedy of the Commons" is a central idea in understanding why humans have taken so long to develop sustainable practices.

External Costs

There are now ways to reduce the Tragedy of the Commons of environmental damage. Understanding how this occurs is based on how *external* costs become *internal* costs as companies manufacture products.

Internal costs are those costs that a company actually experiences in manufacturing a product. For example, a computer would need raw materials, labor, marketing, transportation, and other costs paid by a computer company. These internal costs lead to the pricing of a product, which leads to the decision by the consumer to purchase the product.

However, there may be other costs associated with that computer. For example, who pays the medical bills for the person who gets sick because she lives close to the dump where the toxic waste from computer manufacturing is stored? *External costs* are those costs to people or society that are not directly experienced by the company, and are not usually passed on to the consumer. External costs are felt by someone, however, at some point in time.

There are methods to convert external costs into internal costs. Establishing laws and creating taxes are two ways to internalize external costs. For example, passing the Surface

Mining Control and Reclamation Act now ensures that the price for newly mined coal includes the cost of reclaiming the land after coal mining is completed. In order to begin mining operations, a coal-mining firm must first establish a trust fund that finances the cleanup of the area after mining operations have ceased. The cost of cleanup is factored into whether or not it is cost-effective to undergo the mining operation at that site.

When a company internalizes their external cost of a product, the consumer pays for the full cost of a product or service, or *true cost*. The cost-benefit analysis that involves true costs is sometimes called *full-cost analysis*, or *true-cost pricing*. Full-cost analysis leads to a clearer picture of the marketability of a product and often leads to a cleaner environment because the overall health of the environment is taken into account.

Buying Local

The economics of purchasing from local producers is becoming increasingly important. While it may be less expensive to purchase something that is produced at very high volumes, the transportation costs add to the cost of the product. This can make locally-produced items more attractive economically. This is particularly true because transportation has not only a financial cost, but an environmental cost and sometimes a cost to human health. Consumers who want to minimize the environmental impact of transportation may be willing to pay a little more for a product.

> **DID YOU KNOW?**
> Part of the external cost of oil that is not reflected in the price of gasoline at the pump is the environmental and human cost of retiring old oil tankers. European oil tankers at the end of their useful life are sometimes taken to illegal "ship breaking" harbors, such as the coast of Bangladesh. Here, ships are scuttled and sold for parts. However, the workers are exposed to toxic chemicals and hazardous working conditions because they are desperate for work. The surrounding beaches are also exposed to the same toxic chemicals. Workers and inhabitants pay a price with their health. That is one of the external costs of using products that come from crude oil.

Globalization

With the Internet and efficient transportation, the global economy has become much more tightly interrelated. A single product, such as a computer, may contain parts from many different regions of the world. However, individual countries have different costs of living and environmental regulations. Labor to build those products may shift from country to country. In this situation, there is room for the exploitation of workers (employees who work in environmentally dangerous conditions) and countries that are exploited because they are willing to accept long-term pollution in exchange for cash in the short term. The "commons" is being depleted. All external costs have not been included to create a full-cost analysis of such products.

From an environmental standpoint, it is clear that the pollution of one country affects many other countries and that all countries need to agree on global environmental goals if there is to be any improvement. For example, international cooperation around the Montreal Protocol demonstrated significant improvement in ceasing production of chlorofluorocarbons that contributed to reducing stratospheric ozone. Once again, from the perspective of the Tragedy of the Commons, the atmosphere is part of the commons; these protocols help all the users of the atmosphere be responsible for its cleanliness and improve the ability of other countries to use it as a resource in a sustainable manner.

The World Bank is a financial institution that provides loans to developing countries in the hope of reducing poverty and disease, and increasing environmental sustainability. While the overall objectives of the World Bank appear noble, the institution has been criticized for funding projects that are environmentally shortsighted, or serve the needs of a few, economically powerful countries. Even so, it has helped several developing countries in ways they could not have afforded by themselves.

Relevant Laws

The following laws have not been mentioned previously, but they are important to remember. Think about how the following laws have reduced the Tragedy of the Commons?

The **Montreal Protocol (1987)** sets a timetable for phasing out the use of ozone-depleting substances.

The **Kyoto Protocol (1997)** is an agreement among 150 countries to reduce greenhouse gases.

The **National Environmental Policy Act (NEPA, 1969)** established the Environmental Protection Agency and requires environmental impact statements for major federal construction projects.

The **Occupational Safety and Health Act (OSHA, 1970)** allows employees to hold employers accountable for unhealthy situations in the workplace. This law helps employers internalize potential external health costs of employees.

The **Clean Air Act (1970)** established standards for primary and secondary pollutants and "criteria pollutants" that most threaten human health.

The **National Park Service Act (1916)** created the national park system.

The **Endangered Species Act (1973)** creates a list of endangered species and protects the species and their habitats.

The Lacey Act (1990) prohibits the movement of regulated or poached species across state lines and allows the Department of the Interior to restore populations of scarce or extinct animals or introduce scarce populations to new habitats.

TEST TIP

There are many more environmental laws and treaties than those listed in this chapter. However, for the AP Environmental Science exam, concentrate on being familiar with the pieces of legislation that are reviewed in this chapter.

Time for a quiz
- Review strategies in Chapter 2
- Take Quiz 4 at the REA Study Center
  (www.rea.com/studycenter)

Chapter 8

Energy Resources and Consumption

This chapter covers 10% to 15% of the AP Environmental Science curriculum and is perhaps one of the most critical chapters because the manner by which we obtain energy is related to all other environmental issues. While you may have studied energy in previous science courses, this chapter presumes little background in energy, and portrays energy in a very practical manner.

The College Board syllabus indicates that the following topics from this chapter may show up on the AP Environmental Science exam and will compose about 10% to 15% of the total test, which is in line with the proportion of coverage in the course curriculum.

- A. **Energy Concepts**
  (Energy forms; power; units; conversions; Laws of Thermodynamics)

- B. **Energy Consumption**
  1. History (Industrial Revolution; exponential growth; energy crisis)
  2. Present global energy use
  3. Future energy needs

- C. **Fossil Fuel Resources and Use**
  (Formation of coal, oil, and natural gas; extraction/purification methods; world reserves and global demand; synfuels; environmental advantages/disadvantages of sources)

- D. **Nuclear Energy**
  (Nuclear fission process; nuclear fuel; electricity production; nuclear reactor types; environmental advantages/disadvantages; safety issues; radiation and human health; radioactive wastes; nuclear fusion)

- E. **Hydroelectric Power**
  (Dams; flood control; salmon; silting; other impacts)

F. **Energy Conservation**
(Energy efficiency; CAFE standards; hybrid electric vehicles; mass transit)

G. **Renewable Energy**
(Solar energy; solar electricity; hydrogen fuel cells; biomass; wind energy; small-scale hydroelectric; ocean waves and tidal energy; geothermal; advantages/disadvantages)

Energy Concepts

Energy is the ability to do work. Energy is defined as a force applied over a distance, although it need not necessarily involve the most obvious example of pushing an object; it may also involve a chemical reaction, or the movement of a tectonic plate.

There are two types of energy: **kinetic energy** and **potential energy**. The sum of these two types of energy is called the **total mechanical energy** of an object.

- **Kinetic energy** is the energy that comes from the motion of an object and is proportional to the square of the object's velocity.

- **Potential energy** is stored within the object in one of three ways: gravitational potential energy (as with water stored in a reservoir to be used later for power production), chemical potential energy (such as fats that are stored in our body to be used later, or gasoline that can be used for a car's fuel), and nuclear potential energy (such as in the nucleus of unstable atoms that are used for fuel in a nuclear reactor).

Energy sources for Earth include three different sources. The most common source is the sun, which radiates energy onto the Earth. Two less common sources of energy include the Earth's core, which heats up earthen material and drives plate movement; and nuclear reactions amid unstable elements on Earth, which power nuclear power plants.

The **Laws of Thermodynamics** define two fundamental principles that are at the core of many AP Environmental Science topics. (There is a Third Law of Thermodynamics, but it is not applicable to the AP exam.)

- The **First Law of Thermodynamics** states that the total amount of energy in the universe is constant; it is neither created nor destroyed. This is also called conservation of energy.

- The **Second Law of Thermodynamics** states that when energy changes forms, some energy is lost as waste heat. That is, there is less energy to do work than there was initially because transfers of energy from one form to another are not perfectly efficient. The portion of energy lost due to inefficiency is in a more random form and is called entropy.

TEST TIP

In AP Environmental Science, many of the topics covered in one chapter relate to topics covered in another chapter. The Second Law of Thermodynamics is an important unifying idea in AP Environmental Science. For example, the reason that a food pyramid is shaped like a pyramid is because of the Second Law. The inefficiency of energy moving from one trophic level to another is either because much of the energy is used by the prior trophic level or because random, low-quality heat is lost as entropy at each step. Can you think of other instances where the Second Law of Thermodynamics explains an environmental principle?

Energy Units and Calculations

Energy units can be derived from many types of equations that involve energy. However, the fundamental units for all of these relationships are the same, with the metric units of energy being joules.

$$1.0 \text{ Joule} = 1.0 \text{ kg} \times \frac{m^2}{sec^2}$$

Unit Conversions

Many AP Environmental Science problems involve converting one set of units to another. These problems most often show up in energy topics. For example, a comparison is much easier when one compares numbers that have the same units. To answer these problems, basic factor-label problem solving, as described in Chapter 2, is the easiest and most reliable method to use.

EXAMPLE One pound of coal used by a power plant gives off 5,000 BTUs of heat energy. How many BTUs are produced by 2,500 pounds?

SOLUTION $1.25 \times 10^7 \text{ BTUs} = 2{,}500 \text{ lb coal} \times \dfrac{5{,}000 \text{ BTU}}{1.0 \text{ lb coal}}$

TEST TIP

Energy calculations are much easier if you write down the given information and visualize the number and units together. Then use factor-label problem-solving techniques (Chapter 2) to walk patiently through the problem. What might seem like a hopelessly complex problem at first glance can turn into a simple two-step solution. Don't give up!

Electrical Energy Calculations

Power is the amount of energy exerted in a given time and is expressed in watts.

1 watt = 1 joule/sec

Electrical energy is often expressed in terms of watts, instead of joules, by multiplying both sides of the equation by seconds. This gives us an important relationship when considering problems that relate to electrical energy.

1 joule = 1 watt-sec

EXAMPLE How long will a 100-W bulb shine with an energy input of 1.0 kW-hr?

SOLUTION $10.0 \text{ hr} = 1.0 \text{ kW-hr} \times \dfrac{1{,}000 \text{ W}}{1 \text{ kW}} \times \dfrac{1}{100 \text{ W}}$

EXAMPLE How many joules are in 10.0 kW-hr?

SOLUTION $3.6 \times 10^7 \text{ J} = 10.0 \text{ kW-hr} \times \dfrac{1 \text{ joule}}{\text{W-sec}} \times \dfrac{1{,}000 \text{ W}}{1 \text{ kW}} \times \dfrac{3{,}600 \text{ sec}}{1 \text{ hour}}$

TEST TIP

Don't forget that when you divide by units, you turn them upside down and multiply. For example, to divide by miles-per-hour, you simply multiply by hours-per-mile. With electricity, remember that you might cancel out the units of kW in kW-hr, but you are still left with the units of hours.

Specific Heat Calculations

Specific heat capacity refers to a material's ability to absorb heat. For example, it takes more heat to raise one gram of water by one degree than it does to raise the same amount of sand by one degree. Therefore, water has a higher capacity to hold more heat, or a higher specific heat capacity. The relationship between heat absorbed (or taken away), mass of material, and change in temperature is given by the formula:

$$q = m \, C \Delta T$$

where: q = amount of heat absorbed by the material

m = mass of the material

C = specific heat capacity of the material

ΔT = change in temperature of the material (final to initial)

EXAMPLE How many joules of heat are needed to heat 5.0 grams of water by 10.0°C? (C_{water} = 4.2 J/g · °C)

SOLUTION Use the formula, $q = m \, C \Delta T$.

$210 \text{ J} = 5.0 \text{ g water} \times \dfrac{4.2 \text{ J}}{\text{g} \cdot \text{°C}} \times 10.0 \text{ °C}$

EXAMPLE How much energy does it take for a dishwasher to heat 200 grams of water from 20°C to 90°C? (C_{water} = 4.2 J/g ·°C)

SOLUTION Use the formula, $q = mC\Delta T$.

$$5.9 \times 10^4 \text{ J} = 200 \text{ g water} \times \frac{4.2 \text{ J}}{\text{g} \cdot {}^\circ\text{C}} \times 70.0{}^\circ\text{C}$$

DIDYOUKNOW?

The high specific heat capacity of water is the reason why cities located on the coastlines have a more moderate average temperature than inland cities. When it is warm, the water absorbs more heat to remain at the same temperature. When it is cold, the water releases heat.

Human Energy Consumption

History of Human Energy Use

- **Hunter-Gatherers**

    In prehistoric times, bands of people used fire from wood fuel as a primary source of energy. To do this required that people move from place to place and allow the fuel and food to replenish itself. However, this method of energy use became inadequate as population growth increased.

- **The Neolithic Revolution**

    With the advent of farming and animal domestication about 10,000 years ago, human domestication of energy-use evolved from a wood-burning fire to harnessing muscle energy from animals (to pull carts and plows). Farming allowed a population of people to concentrate its sources of food and grow crops to feed domesticated animals, rather than just hunt wild animals. However, the source of heat energy to cook was still limited to fuel from wood and, sometimes, animal dung. Even today, cultures that rely on Neolithic methods of farming and fuel use are limited by how far they need to travel each day just to get wood for cooking.

- **The Industrial Revolution**

    Four major advances led to the industrial revolution: improved agriculture, the steam engine, the ability to make steel, and the increased use of coal as an energy source. Each of these advances played off each other to allow greater population densities in cities and to improve technology and economic growth.

    Better transportation meant that food could be grown further from high-density population centers. The use of coal brought a source of fuel that was not dependent on nearby forests, the ability to make steel, and the invention of the steam engine (which replaced many uses of animals). In turn, the steam engine helped pump water from coal mines, further improving the supply of coal. People began supporting their families by working in factories, without being dependent upon land to provide food. Greater wealth also brought greater dependence on a non-renewable fuel source and a greater per capital use of energy.

- **The Automobile Society**

    A new energy source, refined crude oil, allowed society to be organized around the automobile. This allowed workers to live farther from work, the market, and their source of food. However, it again stepped up society's dependence on non-renewable energy sources and increased the amount of energy used by the average person. With electrification, more daily chores were made easier, but at the cost of an escalating use of fossil fuels and, eventually, nuclear fuels—yet another non-renewable energy source. In developed countries, people in warm climates began to expect to be able to live in the comfort of an air-conditioned room—further increasing the per capita use of energy from nonrenewable sources.

- **Developed versus Developing Countries**

    Different countries in the world are at different stages of economic development, which coincide with the progression of using different sources of energy. Some developing countries still depend on individuals gathering naturally grown wood in order to cook a meal. Some countries have been able to organize large-scale energy projects, such as hydroelectric dams, nuclear power plants, and clusters of coal-fired power plants to fuel a fully electrified society. Currently, about one-fifth of the world's population lives in a developed country; but those countries use about 80% of the world's energy.

- **The Energy Crisis**

  In the mid-1970s, Western support of Israel during the Yom Kippur War led many Arab nations to place an embargo on selling oil to the Western world. This embargo occurred at a time when developed countries were increasing their use of gasoline and oil-based products at an otherwise unhindered rate. It was common in the United States to experience long lines to fill a car's gas tank. It also coincided with an increased awareness of environmental issues, including the drawbacks of depending heavily on nonrenewable energy sources.

Nonrenewable Energy Sources

The primary method of producing electrical energy from nonrenewable sources is to heat water into steam, which then turns a turbine.

The Formation, Extraction, and Combustion of Fossil Fuels

Coal—About 300 million years ago, during the carboniferous period, conditions on Earth favored freshwater swampy ecosystems. As the plant material from these swamps died, they decayed and became compacted to form *peat*, which is about 5% carbon. Further compaction produces *lignite* coal, which is about 30% carbon. Further compression yielded *bituminous* coal (about 75% carbon). The most compressed form of coal is *anthracite*, which is over 90% carbon.

Coal is the most abundant nonrenewable energy source and is used for about 27% of the world's energy needs. Coal is mined in either an open-pit surface mine or a shaft mine.

Combustion of coal yields oxides of sulfur. Sulfur remains within the coal from decomposed proteins. When combined with water, oxides of sulfur produce acid rain. Acid rain is also produced from oxides of nitrogen, another product of coal combustion.

Coal also contains mercury. The combustion of coal is the largest source of atmospheric mercury. Atmospheric mercury, when inhaled, causes neurological damage.

Oil—In the same manner that coal is formed from plants, crude oil and natural gas are formed by the decomposition of microorganisms. Oil typically exists within the pores of sandstone and shale. Crude oil is a mixture of many hydrocarbons, which can be separated during the refining process. Oil and gas are extracted by drilling wells. These wells may be on land or on offshore rigs that drill into the seafloor.

Natural gas—Natural gas is often located along with crude oil in sedimentary rock, but it can also be found independent of crude oil.

Hydraulic fracturing, or *fracking,* is sometimes used with both oil and gas to increase the yield from a particular site. Fracking involves creating a pathway between rock that contains oil or gas, then pumping water and/or chemicals through the rock to push the oil or gas to a location where it can be retrieved with a well. Fracking increases yields, but many are concerned about pollution of the groundwater and other environmental hazards associated with this new technique.

World reserves for coal are summarized in the following chart. Notice which three countries have the largest coal reserves. How are these reserves typically used?

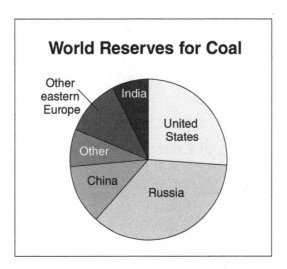

World reserves for oil are summarized in the chart below. How does the United States' reserves for coal differ remarkably from the United States' reserves for oil? How are these two types of energy used differently in our daily life?

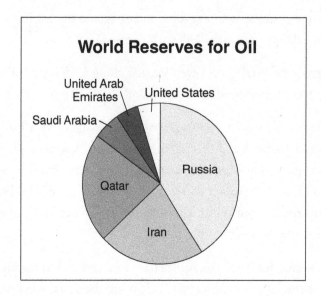

Global demand for energy is easily summarized on the following chart. Why is the use of biomass as high as it is? What types of energy production might qualify to be included in "other"? What percentage of the total energy used comes from fossil fuels?

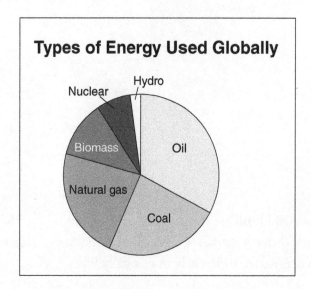

Table 8-1. Pros and Cons of Getting Energy from Fossil Fuels

Pros	Cons
Considerable fuel available in domestic reserves.	Atmospheric pollution that includes acid rain and toxic compounds.
	Mining and drilling—especially with new extraction techniques—causes considerable damage to the land.
	Some say that the combined economic costs, when external costs are included, far exceed some renewable sources.

DIDYOUKNOW?

In the 1930s, a series of locks and dams allowed tug-driven barges to transport coal to a series of 50 coal-fired power plants along the Ohio River. This massive development of the river created an economic boom in Kentucky coal fields, as well as in cities along the eastern seaboard (as a result of the increased availability of inexpensive electricity). However, thermal pollution in the river, acid rain in the air, and heavy metal toxicity have created an environmental nightmare in the region.

Nuclear Energy

Nuclear energy comes from the energy within atomic nuclei. The heat from nuclear reactions boils water into steam, which turns a turbine and produces electricity. To understand how nuclear energy works, you have to review the important nuclear reactions, calculate rates of reactions, understand the types of nuclear reactions that produce electricity, and look at the pros and cons of using nuclear energy.

Nuclear reactions rearrange the structure of individual atoms, often giving off energy in the process. Nuclear reactions are different from chemical reactions in that mass is not superficially conserved; the energy from nuclear reactions comes from mass being converted into energy through Einstein's famed relationship, $E = mc^2$. When these reactions occur in a nuclear power plant, the resulting heat boils water into steam, which turns a turbine, which produces electricity. The most important nuclear reactions have to do with reactions that occur in nuclear power plants, and reactions that occur in the environment.

Fission is the breakup of large, unstable nuclei into many smaller nuclei. This is the nuclear reaction that takes place in power plants, and in atomic weapons. The neutrons that are given off in a fission reaction (see example below) in turn catalyze other large nuclei to split, which in turn results in a chain reaction. This chain reaction can become explosive unless neutrons are absorbed with a control rod, which slows the chain reaction enough so that it emits an amount of heat that is useful in a power plant.

$$^1n + {}^{235}U \rightarrow {}^{92}Kr + {}^{141}Ba + 3\ {}^1n$$

Fusion is the combination of two small nuclei into one larger nucleus, yielding very large amounts of energy and radioactivity. This type of nuclear reaction occurs only at very high temperatures, such as those that occur on the surface of the sun, and with thermonuclear weapons. To date, we have not figured out how to control fusion reactions well enough to generate electricity.

$$^2H + {}^2H \rightarrow {}^4He$$

Alpha decay emits an alpha particle, or the nucleus of a helium atom, and occurs during natural radioactive decay. For example, uranium that is used for nuclear fuel naturally decays into other radioactive nuclei during this method of decay. Radioactive decay of an atom continues until a stable isotope is formed.

$$^{238}U \rightarrow {}^{234}Th + {}^4He$$

Beta decay is another type of natural radioactive decay, which in this case produces an electron. Like alpha decay, it is one mechanism by which naturally radioactive atoms decompose and eventually turn into stable, nonradioactive atoms.

$$^{234}Th \rightarrow {}^{234}Pa + {}^0e$$

Neutron capture is important in environmental science because it is the mechanism by which a nonradioactive atom can become radioactive. For example, during nuclear accidents when a nuclear power plant "melts down," it is through neutron capture that other surrounding elements become radioactive. The neutrons from fission reactions bombard otherwise stable compounds, making them unstable, or radioactive.

$$^1n + {}^{127}I \rightarrow {}^{131}I$$

Rates of nuclear reactions are measured by a radioactive element's half-life, which is the length of time required for half of the atoms in a given sample to disintegrate into the next element along the decay series. Therefore, the actual length of time that a sample of material remains radioactive is equal to the combination of many half-lives.

EXAMPLE If the half-life for a radioactive isotope of iodine is 8 days, what percentage of a sample of this isotope will remain radioactive after 24 days?

SOLUTION Twenty-four days represents three half-lives. During each half-life, the amount of the radioactive isotope decreases by half, or 50%. After the first half-life, 50% remains. After the second half-life, 25% remains (half of 50%). After the third half-life, 12.5% remains (half of 25%). The answer is 12.5%.

EXAMPLE A sample of radioactive material gives off 80 mrems upon being discovered in a cargo transport. The sample was isolated and found to give off 1.25 mrems after six years. What is the approximate half-life of the material?

SOLUTION Another way to solve half-life problems is to use fractions to consider what percent of the original amount is left. After one half-life, ½ is left. After two half lives, ¼; after three, ⅛; and so on. The denominator doubles with each new half-life. The number 1.25 is about 1/64 of the original amount, or six half-lives. To have the substance undergo six half-lives in six years means that it has a half-life of approximately one year.

TEST TIP

Half-life questions can show up in the AP Environmental Science exam. While they require little math at the AP level, they can be tricky. Remember that one half-life cuts the amount of radioactive isotope in half. It does not mean that it all goes away in two half-lives. In fact, it means that, even though you keep having smaller and smaller amounts with each successive half-life, you will continue to have radioactive material for a time equal to many, many half-lives.

Nuclear reactors of all types have some similar characteristics. All reactors bundle fuel together with fuel rods, use control rods to moderate the rate of reaction, require some type of coolant, and turn a turbine to produce electricity.

Photo credit: iStockphoto/Thinkstock

Figure 8-1. Nuclear Reactor

Types of nuclear reactors include the following:

- **Boiling water reactors** allow water to come in direct contact with the fuel assembly and drive the turbine directly. Because this water becomes radioactive and creates a radioactive turbine, this is the dirtiest type of reactor with the greatest chance of having an accident.

- **Pressurized water reactors** allow water to circulate through the core of the reactor to absorb heat and cool the fuel assembly. This water heats a secondary circuit of water, which evaporates to form steam, which turns a turbine to generate electricity.

- **Heavy water reactors** use radioactive water as both coolant and moderator.

- **Graphite reactors** use carbon as both moderator and cooling agent. Then they use gas to carry the heat from the reactor to steam generators, which in turn drive the turbines. The reactor at Chernobyl, located in the Ukraine, which underwent a severe nuclear accident in 1986, was a graphite reactor.

Radiation and Human Health

Radioactivity is energy emitted from nuclear reactions. Like other types of energy, it can damage humans. Humans are constantly subjected to background radioactivity due to cosmic rays, the sun, building materials, and other sources. Usually our cellular machinery has evolved to be able to handle radioactivity at these low doses. However, nuclear accidents and weapons can expose humans to much higher levels of radiation.

Measuring Radiation Doses

The metric units of energy are joules, as described earlier in this chapter. The units of energy for radiation are *rads* (radiation absorbed dose). One *rad* equals 1×10^{-2} (0.01) joules per kg of absorbing material.

The amount of biological tissue damage is related to both energy, measured in rads, and the type of subatomic particle colliding with the body. Biological risk is measured in *rems* (roentgen equivalent for one man). One *rem* equals the number of rads multiplied by a constant that is related to the type of radioactive particle that causes the radiation. The following chart summarizes the causes and effects of various doses of radioactivity, expressed in rems.

Table 8-2. Causes and Effects of Radiation by Dose

Dose (mrem)	Cause	Effect
3 mrem	Five-hour plane flight	None
7 mrem/yr	Building materials	None
50 mrem	Diagnostic X-ray	None
1,000 mrem/yr	Safety threshold	None
10,000 mrem/yr		Cancer risk
30,000 mrem		Decreased white blood cells
350,000 mrem		Half die in 30 days

Radioactive Wastes

Unlike power plants that use fossil fuels, nuclear power plants do not produce air pollution on a daily basis. However, every step of the mining, refining, processing, use, and storage of nuclear fuel creates waste that is radioactive, and represents a possible route for radioactivity to enter the environment.

Originally, advocates of nuclear power envisioned a nuclear fuel cycle where depleted nuclear waste could be reprocessed and used again. In fact, this happens very little. In the United States in particular, nuclear fuel is not reused. We are still trying to figure out how to store it permanently. Waste from power plants is typically stored at the power plant, often within concrete containers or submerged in pools of water.

Table 8-3. Pros and Cons of Nuclear Energy

Pros	Cons
No atmospheric pollution with normal operation.	Harmful radioactive waste produced at every level.
Considerable fuel available in domestic reserves.	Thermal pollution of water, which is needed to cool most power plants.
	Some say that the combined economic costs are larger than the benefits gained from using nuclear power.
	The human cost of an accident at a nuclear power plant is profound. More than 5 million people are considered to have been contaminated by the Chernobyl disaster in 1986.
	Within the sequence of events that produces nuclear fuel, there is the possibility for the proliferation of fuel that could be used as material for weapons.

DID YOU KNOW?

The nuclear fuel "cycle" is not really a complete cycle because much of the radioactive waste from nuclear reactors has no final resting place. Instead it resides either at reactors, or at "temporary" sites, such as the Idaho National Engineering Laboratory outside Idaho Falls. There was a strong effort to create a final resting spot for nuclear waste at Yucca Mountain, which is thought to be a geologically stable location near a former nuclear test site in Nevada. However, this option was eventually blocked by concerned voters.

Renewable Energy Sources

Renewable energy sources can be regenerated within our lifetime at a rate that exceeds its use. With increased pollution and political tension as a major result of fossil fuels, renewable energy sources suggest a possible way to a cleaner, healthier, safer future. Energy sources that are considered renewable include solar, hydro, biomass, wind, tidal, and geothermal.

Solar Energy

Solar energy is the ultimate source of many, but not all, renewable and nonrenewable energy sources. For example, wind energy ultimately comes from solar heating of the atmosphere. Oil ultimately comes from solar energy reaching photosynthetic organisms. By contrast, tidal energy comes from the potential energy of the moon acting upon the Earth. Geothermal energy comes from the heat from the Earth's core. Energy from the sun can be divided into two categories, passive solar energy and active solar energy.

- **Passive solar energy** uses design of orientation, special materials, and space to maximize the retention and flow of solar energy. Passive solar energy does not involve an input of electricity or mechanical advantage to make it work.

  Building design is a good example of capturing passive solar energy. Some design elements that involve passive solar energy include the following:

  — **Trombe walls**, a massive wall built behind a window that is exposed to the sun to retain and re-emit solar heat;

  — **Extended eves** that block summer sun and allow winter sun to enter a window;

  — **Greenhouses** with vents that allow solar-heated air to enter a living space.

  Planting deciduous trees blocks the summer sun and allows the winter sun to pass through and heat a space.

- **Active solar energy** refers to the input of energy in order to gain full benefit from solar energy, or the direct conversion of solar energy into electricity.

  Heating water is probably the most efficient type of solar heating, where solar gain is focused onto water, or other heat conductor, to circulate and heat spaces, or heat water directly.

Photovoltaics are made of semiconductor material, and use photons directly to produce an electrical current. The typical 2 × 5 foot photovoltaic panel produces electricity at a rate of about 230 watts in full sun. They can be mounted to buildings or built into large arrays.

Hydrogen fuel cells can be used to store energy gained from the sun, so that the energy can be used when the sun is down. They are not used to generate energy. Hydrogen fuel cells use electricity from a source such as photovoltaic panels, wind turbines and other sources of energy, to obtain hydrogen and oxygen from water molecules. At a later time, when the energy is needed, the hydrogen and oxygen come back together to form water, and the electricity is returned with high efficiency and no pollution.

Drawbacks of solar energy include solid-waste challenges when photovoltaics reach the end of their effective life span.

Hydroelectric Power

Hydroelectric power is typically generated with dams that impound water, forming a lake or reservoir behind the dam. The water then flows in a controlled manner either over a spillway, or through turbines that turn generators that produce electric power.

Small-scale hydroelectric units are able to divert the flow of small streams toward a hand-sized water turbine, which generates household levels of electricity. "Micro-hydro," as it is sometimes called, is a good solution for remote residences, especially when used in combination with other renewable energy sources, such as active and passive solar, and wind.

Table 8-4. Pros and Cons of Hydroelectric Power

Pros	Cons
Highly efficient production of electricity	Impounded water devastates habitat
No toxic by-products or pollution	Siltation behind dam, prevents the transport of fertile nutrients downstream
Renewable energy source	River is blocked, reduces fish migration
Transportation costs minimal	Impounded water displaces people
Predictable water supply and flood control	Impounded water increases evaporation
Recreation in lake behind dam	Reduced fertility of flood plains

Biomass

Biomass energy refers to the energy from living tissue—usually bacteria or plants. The following are important components in understanding this type of energy:

- **Plants convert solar energy into chemical energy** stored inside the bonds of organic molecules. When those chemical bonds are broken—such as when we burn wood—most of the energy stored in the bonds is released.

- **Fermentation** is another way to capture solar energy. Yeasts metabolize sugars to produce alcohol, which can be used later as fuel.

- **Bacterial digestion** of carbon wastes—including animal waste and trash—can also produce methane, or natural gas, which can be burned to provide heat or run a generator to produce electricity.

- **Biodiesel** is fuel that comes from the oils found in plants, such as vegetable oil, soybean oil, and sunflower oil. Cars with diesel engines can be powered directly on these natural oils.

Drawbacks of biomass energy:

- **Deforestation** in developing countries is one of the drawbacks of using biomass as a primary energy source.

- **Air pollution** through the production of ash and greenhouse gases is a downside of using biomass as a fuel for combustion.

Wind Energy

About 2% of the solar energy that strikes the Earth heats the air and creates wind, which carries the ability to do mechanical work. Wind generators are 35% to 59% efficient in converting mechanical energy into electrical energy. While they require considerable maintenance, they produce no emissions. The use of wind varies, from thousands of small windmills in rural areas to pump water, to large wind farms that use aerodynamically designed blades and high-tech rotors to generate electricity.

Photo credit: iStockphoto/Thinkstock

Figure 8-2. A Wind Farm

Tidal Energy

Tidal energy ultimately comes from the moon's gravitational action on the oceans, and the tidal movement that results. Like wind and solar energy, the electricity generated from tidal movement is intermittent. Like capturing hydroelectric power, capturing tidal energy can mean ecological disruption in those areas where a high tide is retained, and then allowed back to the sea by moving downhill through an electrical generation station.

Geothermal Energy

There are two types of geothermal energy that can be tapped:

Hot water from mantle heating can drive a generator to produce electricity. This type of geothermal energy can be used only in those rare areas where there are thermal features, such as hot springs, near the surface of the Earth.

Earth heating and cooling using a heat pump is a type of geothermal energy that is often neglected. A heat pump is the type of mechanism used in a refrigerator to isolate heat.

The same mechanism can be used to either pump heat into the Earth (cooling) or draw heat out of the Earth (heating). The Earth is used as a nearly limitless heat sink. This type of geothermal energy can be used any place on Earth.

Energy Conservation

The conservation of energy can take place in a number of different types of human activities. First, however, it is important to understand the concept of energy efficiency and how that relates to conserving energy.

Energy efficiency is important for minimizing our use of energy. If we can be more efficient in producing and using energy, then the resources that go into producing energy can be saved. Before understanding what types of behaviors can more efficiently use energy, let's take a look at how to calculate energy efficiency.

Energy efficiency can be measured by the amount of energy available to do work, divided by the total amount of energy produced. The difference between these two amounts of energy is the amount of heat that is lost due to entropy, as described by the Second Law of Thermodynamics.

$$\text{Energy efficiency} = \frac{\text{energy available to do work}}{\text{total energy produced}} \times 100$$

EXAMPLE A coal-fired power plant converts chemical energy stored in coal into electrical energy through many different steps. The combination of steps carries an overall efficiency of 30%. What amount of electrical energy, expressed in kW-hr, is available to do work for each MW-hr worth of coal that is burned?

SOLUTION $\text{Energy efficiency} = \frac{\text{energy available to do work}}{\text{total energy produced}} \times 100$

Energy available to do work = efficiency × total energy produced

$$300 \text{ kW-hr} = .30 \times 1 \text{ MW-hr} \times \frac{1000 \text{ kW}}{1 \text{ MW}}$$

EXAMPLE A microhydro generator transfers energy from the movement of water into an electrical current using multiple transfers of energy. The moving water turns a turbine that is 70% efficient. The turbine turns a group of magnets, which loses about 15% of the captured energy as heat. The turning magnets induce an electrical current in nearby copper wires, a process that is 60% efficient. What is the overall efficiency of the microhydro generator?

SOLUTION The combined efficiency of the generator is found by multiplying the efficiencies for each step together. Remember, for the second step, the efficiency equals 100% – heat lost.

Total efficiency = $0.70 \times 0.85 \times 0.60 = .36$, or 36% efficient

Energy efficient lifestyle choices can make a large impact on how each of us uses energy. The lowest energy use results from a combination of reducing energy use, using products and practices that are more efficient in using energy, and using products and practices that are sustainable over time. The following are a number of lifestyle choices that affect the total amount of energy we use:

- **Recycling** can reduce our overall use of energy. For example, metal that comes from a recycled aluminum can needs about 5% of the energy that it would take to make a new aluminum can from scratch.

- **Living closer to markets and work** means using less gasoline to commute because we don't need to drive as far, or even drive at all.

- **Hybrid cars** use less gasoline. Hybrid cars have two engines: a regular internal combustion engine and an electric engine/generator. Going downhill and braking causes the generator to produce electricity, which is either stored in batteries or used to propel the car. When the electrical energy is used, less gas needs to be burned by the combustion engine.

- **Public transportation** uses far less fuel because more people are served by the use of a single combustion engine. Even less pollution results if the public transportation is electric, and that electricity is supplied by a clean, renewable source, such as solar energy.

- **Low-energy consumerism** might include purchasing products made closer to the point of use, so that less energy is used to transport them; or with less packaging, so that less energy is used to manufacture needless packaging.

- **Low-energy homes** are increasingly more common. A home uses less energy if it is well insulated and sealed and if it is outfitted with appliances certified by "Energy Star." LEED-certified homes, a certification available through the United States Green Building Council, use less energy to build, maintain, and operate.

- **Low-energy recreation** might include a hike at a nearby lake, versus using a trailer to transport an "all terrain vehicle" to use on trails many miles away.

Corporate Average Fuel Economy (CAFE) standards were first enacted into U.S. law in 1975 and updated in 2011. CAFE standards combine fuel economy and the weight and size of a vehicle, so that larger vehicles require less fuel economy. These standards were modified in 2007 with the Energy Independence and Security Act, which set an overall goal of 35 mpg for all vehicles by 2020. Proponents of CAFE assert that the overall use of fuel has been less with that policy. Opponents of CAFE assert that the policy has let larger vehicles off the hook with respect to environmental responsibility.

Time for a quiz
- Review strategies in Chapter 2
- Take Quiz 5 at the REA Study Center
  (www.rea.com/studycenter)

Pollution

Chapter 9

The College Board syllabus indicates that the following topics from this chapter may show up on the AP Environmental Science exam. This chapter contributes more to the AP Environmental Science exam as these topics compose about 25% to 30% of the total test.

A. **Pollution Types**
   1. Air pollution (Sources—primary and secondary; major air pollutants; measurement units; smog; acid deposition—causes and effects; heat islands and temperature inversions; indoor air pollution; remediation and reduction strategies; Clean Air Act and other relevant laws)
   2. Noise pollution (Sources; effects; control measures)
   3. Water pollution (Types; sources, causes, and effects; cultural eutrophication; groundwater pollution; maintaining water quality; water purification; sewage treatment/septic systems; Clean Water Act and other relevant laws)
   4. Solid waste (Types; disposal; reduction)

B. **Impacts on the Environment and Human Health**
   1. Hazards to human health (Environmental risk analysis; acute and chronic effects; dose-response relationships; air pollutants; smoking and other risks)
   2. Hazardous chemicals in the environment (Types of hazardous waste; treatment/disposal of hazardous waste; cleanup of contaminated sites; biomagnification; relevant laws)

C. **Economic Impacts**
   (Cost-benefit analysis; externalities; marginal costs; sustainability)

Air Pollution

Sources of Pollution

Natural sources contribute air pollutants such as dust from wind erosion, volcanic debris, and sulfur compounds from natural sources. Additionally, volcanoes give off an acidic plume mixed with caustic particulate matter. Trees and bushes can give off volatile organic compounds. Many people suffer from the natural emission of pollen. Forest fires give off extraordinary levels of carbon dioxide and particulate matter. Some rocks in the ground give off radioactive radon. These natural sources combine with *anthropogenic*—or human-caused—sources to create the total impact on human health.

Anthropogenic sources can be divided into the following categories:

1. **Primary pollutants** are harmful to humans in the form in which they are originally released.

2. **Secondary pollutants** are released in a form that is initially not harmful, but they become toxic or hazardous after they are released. Photochemical smog is an example of a secondary pollutant that becomes damaging only after sunlight catalyzes the formation of oxidants and acids.

3. **Fugitive emissions** do not come from a single source, or smokestack, but from a number of nonlocalized sources. Examples include dust from wind erosion of soil or mining operations, auto emissions, or volatile organic compounds from residential backyard grills. Fugitive emissions that are pollutants can either be primary pollutants or secondary pollutants.

The chart on the next page summarizes the general sources of air pollutants—notice which sectors represent the top three sources of air pollution.

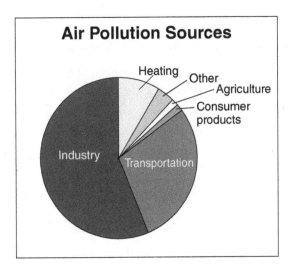

TEST TIP

This chapter represents the largest contribution to the AP Environmental Science exam over all the other chapters, and yet there is not that much information here relative to some other chapters. This chapter uses charts to summarize information whenever possible. Be sure to know each detail in this chapter!

Major Air Pollutants

Noncriteria pollutants are not regulated and do not require monitoring by the federal government. However, just because the government does not regulate a pollutant does not mean that it is not harmful. Indoor air pollutants, mentioned later, is one example of noncriteria pollutants that can have a severe effect on human health. However, in this chapter, let's spend our time reviewing criteria pollutants.

Criteria pollutants are regulated by the federal government. Initiated by the Clean Air Act of 1970, this legislation identified seven major air pollutants as being the most damaging to human health, and established standards for those pollutants. Another name for criteria pollutants is conventional pollutants.

1. **Oxides of Nitrogen**

   Description: Oxides of nitrogen are formed when nitrogen from the atmosphere is combusted at high temperatures, above 650°C. This results in oxygen in the atmosphere combining with the nitrogen to form oxides of nitrogen. Nitrite and nitrate ions are formed when bacteria oxidize nitrogen from the atmosphere or ammonia as part of the nitrogen cycle. Nitrogen oxides and ions react with water to form nitrous acid and nitric acid, respectively—two major contributors to acid rain.

   Sources: Over half the nitrogen oxides are anthropogenic, mostly coming from power generation and internal combustion engines in cars, trucks, planes, and trains.

   Effects: Acid rain from the hydrated oxides of nitrogen damage waterways and wildlife, forests and plants, buildings, and people. For those who live in mountainous European towns, acid rain destruction of forests translates into soil erosion that threatens hillside villages. Oxides of nitrogen are responsible for about 30% of the acid deposition that occurs.

   Nitrate and nitrite ions are essential nutrients for plants. These nutrients promote photosynthesis in algae, which is then used as food by oxygen-demanding decomposing organisms. The overall effect of these nutrients entering a waterway is the reduction of dissolved oxygen in the water and an increase in eutrophication. For this reason, nutrients are considered *oxygen-demanding wastes*.

   Remediation: Reduction is the best method for preventing pollution from oxides of nitrogen, but the next best method is to avoid getting combustion reactions hot enough to produce the oxides, or use pure oxygen rather than air to fuel combustion reactions.

   In power plants, one way to run combustion reactions at a lower temperature is to use a staged burner. In the first stage, combustion takes place at high temperatures with little oxygen. In the second stage, combustion takes place in an oxygen-rich, low-temperature environment. Both stages result in few oxides of nitrogen. Catalytic converters in cars also reduce nitrogen oxides, as well as carbon monoxide.

2. **Oxides of Sulfur**

   Description: Sulfur dioxide is a corrosive gas that damages the tissues of plants and animals. It can react with water to form sulfurous and sulfuric acid, which is a major component of acid rain.

   Sources: Sulfur dioxide is produced primarily by electric utilities in the combustion of coal. The coal, which is primarily carbon, comes from the organic breakdown of plant material, which contains proteins. Proteins contain sulfur from the disulfide linkages that shape the protein molecule. The protein that is trapped with other organic materials decomposes and leaves some type of sulfur compound within the coal. When the coal is burned, oxygen combines with the sulfur in the coal to form sulfur dioxide, which forms sulfurous or sulfuric acid when combined with water vapor in the atmosphere. Nonanthropogenic sulfur is also emitted from volcanoes.

Effects: Sulfur dioxide causes breathing difficulties. Sulfates combine with other ions to form insoluble particulate matter that irritates the lungs. Sulfur dioxide combines with water vapor to produce acid rain. Oxides of sulfur are responsible for about 70% of the atmospheric acid deposition.

Remediation: Using low-sulfur coal makes a large difference in reducing the amount of sulfur oxides emitted from power plants. Other techniques to reduce oxides of sulfur include using a *fluidized bed combustion*, which mixes crushed limestone with the coal, a mixture that causes the sulfur dioxide to combine with the calcium to form a solid waste rather than atmospheric pollution. These same principles are behind *flue gas desulfurization* and *wet scrubbing* technologies sometimes used in power plants.

3. **Oxides of Carbon**

    Description: Carbon dioxide and carbon monoxide are colorless, odorless gases.

    Sources: Carbon dioxide is used as a reactant in photosynthesis and is a major product in the combustion of any carbon-based fuel, such as wood and fossil fuels. Carbon monoxide comes from the incomplete combustion of wood and fossil fuels.

    Effects: Carbon monoxide is acutely poisonous because it binds more readily to oxygen-carrying hemoglobin molecules in our blood, thereby preventing the transport of oxygen to tissues. Carbon dioxide is also deadly in that it can fool the body into needing to breathe, thereby causing asphyxiation. Carbon dioxide is a greenhouse gas that contributes to climate change, and the subsequent change in ecosystems and environmental calamity described in the last chapter of this book.

    Remediation: Homes should be equipped with carbon monoxide detectors to alert people of its presence. Another issue is how humans will prevent climate change that is happening because of increased carbon dioxide in the atmosphere. Some possible solutions include carbon sequestration with forests, and reducing the use of fossil and wood fuels.

4. **Volatile Organic Compounds (VOCs)**

    Description: VOCs are organic compounds that are easily vaporized, or have a high vapor pressure under normal conditions. Also, VOCs react to form ground-level ozone, which is a strong irritant.

Sources: VOCs arise from evaporation of crude oil, air conditioning fluids, dry-cleaning solvents, paints, adhesives, building materials, and other hydrocarbon sources.

Effects: Of the hundreds of VOCs that are used and found in the environment, different compounds have different effects. Some VOCs are carcinogens, some are irritants, some are neurotoxins, and some cause liver or kidney failure. VOCs are the largest category of the most highly toxic criteria pollutants.

Remediation: Evaporation of VOCs decreases when the sources of the hydrocarbons are isolated so that gas may not escape. Automobile carburetors can be adjusted so that less gasoline evaporates. Fewer VOCs will be emitted into the atmosphere if afterburners are installed so that a more complete combustion takes place and completely oxidizes the hydrocarbon to form carbon dioxide and water. However, these afterburners require very high temperatures. The trade-off with fewer VOCs in this way is the production of more oxides of nitrogen.

5. **Particulate Matter**

    Description: Another name for particulate matter is aerosol, which is a group of solid or liquid particles that are suspended in the atmosphere. *Suspended* refers to the fact that the particles are small enough to be held up by the kinetic energy of the surrounding gas molecules. Particulate matter includes ash, such as from a forest fire, or fly ash from coal-fired power plants, dust, smoke, pollen, and mildew spores.

    Sources: Nonanthropogenic sources include volcanoes, forest fires, leaf mildew, and pollen. Anthropogenic sources include power plants, factories, and a wide range of industrial processes.

    Effects: Of course, particulate matter reduces visibility, but it also irritates the lungs and carries the same toxic traits that the individual solid particles would have. However, the effects are magnified because the material is carried deep into the lungs, where it can diffuse into the blood or react with sensitive lung tissue.

    Remediation: Particulate matter can be removed from power plant emissions by using bag filters or electrostatic precipitators. Bag filters are similar to a vacuum cleaner filter, where flue gas passes through a filter with holes smaller than the particles. The filter catches the particles, and they are later disposed of as solid waste. Electrostatic precipitators add a charge to particulate material in the stack, then the gas is surrounded by a charged plate that attracts the charged particles to it and removes them from the air.

6. **Metals and Halogens**

    Description: Metals enter the atmosphere as volatile gases (as with mercury), oxides, or particulate solids. Metals that pose the greatest concern are lead and mercury, both of which are neurotoxic.

    Halogens are nonmetals that are very reactive, such as chlorine, fluorine, bromine, and iodine. Chlorinated and fluorinated hydrocarbons are particularly important because they migrate to the stratosphere and catalyze the conversion of atmospheric ozone into molecular oxygen, which is transparent to mutagenic ultraviolet (UV) rays. This process destroys the ozone layer that has been built up over many millennia and provides life on Earth with protection from damaging UV radiation.

    Sources: Lead in the atmosphere typically comes from gasoline, to which it has been added to reduce "knocking" and to catalyze more complete combustion. Mercury is emitted from coal-fired power plants and, to a smaller degree, from waste incinerators. Mercury switches in home thermostats may emit low levels of mercury, but they are being replaced by electronic switches.

    Halogens are most often emitted as chlorofluorocarbons (CFCs) from propellants, coolants, foams, and dry-cleaning solvents.

    Effects: Lead and mercury lead to decreased neurological functioning. Lead may decrease the body's ability to metabolize food and lead to nausea. In excess, both metals may lead to permanent debilitation and death.

    The problem with any level of mercury in the environment is that it is metabolized by microorganisms to form highly toxic methyl mercury, which is a potent neurotoxin. Because methyl mercury is fat-soluble and persistent, it tends to undergo both bioaccumulation and biomagnifications (see descriptions later in this chapter).

    The decrease in ozone protection catalyzed by CFCs leads to skin cancer in the short term, but may have a drastic effect on global biology if allowed to continue.

    Remediation: Lead levels have decreased with the use of unleaded gasoline. However, the use of leaded fuels continues at some levels, and so does the problem. Stopping the use of leaded fuels entirely will eliminate atmospheric lead.

    Some mercury is removed from fly ash by electrostatic precipitation, but then it becomes a solid waste problem. Because most of the anthropogenic mercury comes from coal-fired power plants, we will continue to release mercury in the environment as long as we use coal as a fuel to produce electricity.

7. **Photochemical Oxidants**

   Description: Photochemical oxidants are secondary pollutants that are synthesized with the aid of solar energy. The most common of the photochemical oxidants is ozone, which is produced when a single atom of oxygen is split off the nitrogen dioxide molecule. The single atom of oxygen then combines with oxygen gas to form ozone.

   Sources: Because the oxides of nitrogen are so critical to the creation of ozone, the sources and remediation of ozone and other photochemical oxidants are the same as those mentioned regarding the oxides of nitrogen.

   Effects: While ozone is an essential protecting molecule when it exists in the stratosphere, it is a caustic irritant when it is in the troposphere.

Table 9-1. Sources, Effects, and Remediation Measures of 7 Major Air Pollutants

Criteria Pollutant	Sources	Effects on Humans	Remediation
Oxides of nitrogen	Power plants, cars	Acid rain, nutrient pollution	Staged burners, catalytic converters
Oxides of sulfur	Power plants	Acid rain, irritant	Low sulfur coal, fluidized bed combustion, flue gas desulfurization, wet scrubbing
Oxides of carbon	Combustion of fossil and wood fuels	Death; generalized results of climate change	Reduce the use of fossil fuels; catalytic converters in cars; carbon sequestration in forests
Volatile organic compounds	Organic compounds, oil, building products, solvents	Carcinogenic, neurotoxic, liver and kidney toxic, irritants	Reduce the use of organic compounds, respirators, afterburners; solution depends on type of VOC
Particulate matter	Power plants, factories, industries, volcanoes	Toxins vary and are easily absorbed through the lungs.	Filtration; bag filters and electrostatic precipitation in power plants
Metals and halogens	Gasoline, power plants, propellants	Corrosives, neurotoxins	Reduced use
Photochemical oxidants	Same as for oxides of nitrogen	Irritant	Same as for oxides of nitrogen

Indoor Air Pollution

Indoor air pollution can be caused by a number of products used in home construction. A few of the more common toxic materials used in building homes, as well as naturally occurring toxins in the home, include the following:

- **Volatile organic compounds (VOCs)** are compounds that contain one or more carbon atoms. Formaldehyde is probably the most common VOC found in homes. Others include toluene, xylene, and turpentine. VOCs can be found in carpets, carpet padding, caulking, adhesives, particle board, plywood, and insulation. VOCs cause dizziness, nausea, fatigue, headaches, shortness of breath, dermatitis, and even cancer in some cases.

- **Chromium salts** are contained in "green-treated" waterproofed lumber. The chromium portion of this biocide may contain forms that are carcinogenic.

- **Fiberglass microfibers** found in insulation are irritants and suspected of being a mutagen in lung tissue.

- **Asbestos** may be a principal ingredient in old flooring, adhesives, ceiling tile, siding, and other products.

- **Radon** is a naturally occurring radioactive toxin that causes lung cancer, and may be found coming from concrete and granite in building materials, or seeping from natural sources into the basement of homes. It is a by-product of the decomposition of uranium, which is occurring in small amounts throughout the Earth. So any Earth materials may contain radon-emitting particles. Ventilation is the best way to reduce radon in the home.

- **Carbon monoxide** is a colorless, odorless gas that competes with oxygen for the critical binding site on the hemoglobin molecule in red blood cells. If carbon monoxide is present, our tissues can't get oxygen. It is very lethal. Carbon monoxide is created from incomplete combustion in a fuel-burning appliance, such as a furnace, fireplace, grill, and any gas appliance. Every home should contain a carbon monoxide detector.

- **Lead** is a strong neurotoxin and is found in older plumbing and paint.

Table 9-2. Physiological Effects and Sources of Indoor Air Pollutants

Indoor Pollutant	Physiological Effects	Sources
Radon	Lung cancer	Cement, granite, underground rocks
Carbon monoxide	Asphyxiation	Incomplete combustion
Volatile organic compounds	Dizziness, nausea, cancer	Carpets, stain, glue, caulking, insulation foam, plywood
Asbestos	Lung cancer	Old tiling, glue, drop ceilings
Mold	Immune suppression	Any dark, damp space
Cigarette smoke	Headache, cancer	Drapes, carpet, wood
Fiberglass	Irritant, possibly cancer	Insulation
Lead	Neurotoxin, learning disorders	Older paint
Mercury	Neurotoxin, neurological disorders	Older thermostats
Chromium	Cancer, liver/kidney toxin	Pressure-treated lumber

Measuring Air Pollution

Measuring air pollution is typically done using units of micrograms per cubic meter ($\mu g/m^3$), parts per million (ppm), or parts per billion (ppb). Conversions among these units depend on the molar mass of the substance. The following chart shows how 1.0 $\mu g/m^3$ of different substances can have different ppm and ppb values, as well as their air quality standard. Using this chart, can you convert among the three different types of units for these three gases? Which has the lowest concentration for its standard, sulfur dioxide or ozone?

Table 9-3. Standards for Measuring Air Pollution

Substance	Formula	Mol. Weight	$\mu g/m^3$	ppm	ppb	Standard
Nitrogen dioxide	NO_2	46	1.0	0.00050	0.50	100 ppb
Sulfur dioxide	SO_2	64	1.0	0.00077	0.77	75 ppb
Ozone	O_3	48	1.0	0.00048	0.48	0.75 ppm

TEST TIP

Although conversion among the units for measuring air pollution is not typically part of the AP Environmental Science exam, it's helpful to have some knowledge of how it is accomplished.

Effects of Pollutants on Humans

General Effects

- Fifty thousand Americans die prematurely each year from air pollution-related illnesses.

- Residents of polluted cities are 15 to 17 times more likely to die from air-pollution-related illnesses.

- In 2004, the Environmental Protection Agency reported that 159 million Americans breathe air that is unhealthy.

- Half of all autopsies demonstrate some degree of lung degeneration that is typically caused by air pollution.

- Ailments such as heart attacks and immunological disorders are more likely to occur in people who breathe polluted air.

- Worldwide, at least 1.3 billion people live in dangerously polluted areas.

Lung Irritation

- Irritation is caused by strong oxidizers in air pollution, such as sulfur dioxide, nitrogen dioxide, suspended particulate matter, and ozone.

- Suspended particulates penetrate deep into the lungs and can trigger an inflammatory response and asthma. Asthma is usually caused by an allergic reaction—often to fine particulate matter. During an asthma attack, the bronchi fill with mucus and restrict breathing. Thousands die each year as a result of severe asthma attacks.

- Carbon monoxide blocks the ability of red blood cells to carry oxygen to tissues.

- Bronchitis, or inflammation of the bronchial tubes, leads to mucus, coughing, and infection. Bronchitis can lead to emphysema, which breaks down the walls of alveolar sacs and decreases lung capacity.

- Smoking is the largest cause of lung disease. About 3 million people die each year from tobacco-related illnesses.

Asthma Triggers

Asthma affects millions of Americans, including 5 million children. It is the leading cause of childhood hospitalizations and school absenteeism. The EPA identifies secondhand smoke, dust mites, pets, molds, and cockroaches as the most prevalent household triggers of asthma.

Secondhand smoke contains over 40 substances that are linked to cancer, and it is thought to trigger asthma because it irritates bronchial passages. Children are particularly vulnerable to secondhand smoke because they are still developing and have higher respiratory rates than adults.

Dust mites, pets, molds, and cockroaches act as allergens that trigger asthma. Dust mites feed on skin flakes that remain on mattresses and fabric-covered items. The body parts and feces from both dust mites and cockroaches are strong allergens, as is the dander from cats and dogs.

Molds produce microscopic spores that circulate throughout homes. When the spores find a moist environment, they grow more mold. Controlling moisture is the best way to control the level of mold or mold spores in a home.

Effects of Pollutants on Ecosystems

Acid Rain

Effects on aquatic ecosystems: Acid rain causes leaching of metals from substrate, such as aluminum, mercury, and cadmium. These metals can exceed tolerance levels for aquatic organisms. Acid rain changes the pH of the aquatic environment, which may exceed the tolerance limits for aquatic species. A shift in the pH of a stream or lake favors different organisms.

Effects on plants and soils: As with aquatic organisms, a decrease in soil pH increases the leaching of metals out of nearby substrate, which can have a deleterious effect on plant growth. Aluminum ions are particularly destructive for plants. Nutrients essential for plant growth can leach out of the soil and enter the groundwater, thus depleting these nutrients for further plant growth. Also, a change in soil pH changes a major abiotic factor that shifts a soil's ecosystem to favor other plants, such as mosses, which remove air from soil and decrease soil health for other plants. Finally, sensitive cells in plants—such as those in the leaves—are damaged from the caustic action of acids, or they can become more susceptible to disease and pathogens as a result of being exposed to acid.

The Greenhouse Effect

The greenhouse effect refers to the warming of the Earth that occurs when gas molecules absorb the low-energy infrared radiation that has been re-emitted by the Earth. The gases that cause such an effect are called greenhouse gases, and they include carbon dioxide, methane, CFCs, nitrous oxide, water vapor, and a few other trace gases. Of these, carbon dioxide is by far the largest culprit. The largest anthropogenic source of carbon dioxide is the combustion of fossil fuels. (See a more detailed description in Chapter 10.)

Aerosol Effect

Particulate matter, or aerosols from forest fires, urban pollution, and volcanoes, have the opposite effect of greenhouse gases. They actually block the sun's rays and cool the Earth. For example, in 1815, the Indonesian volcano Tambora erupted. Approximately 100,000 people died as a result of the initial blast and also as a result of the subsequent famine after fields were destroyed from ash and acid rain. However, the long-term effects caused cooling on a global scale, decreasing average temperatures by about 3°C. The decrease in available sun-shortened growing seasons and caused other famines in North America and Europe. Many scientists feel confident that sudden cooling was the reason for the disappearance of the dinosaurs. One theory suggest that a massive asteroid struck the Earth and kicked up enough particulate matter to have an effect similar to that which occurred from the Tambora eruption.

Ozone Depletion

First reported in 1985, the periodic creation of a hole in the ozone layer in the stratosphere was startling news. The ozone layer provides a shield for all living organisms from UV light that would otherwise cause burns, cancer, and genetic damage. Ozone tends to be depleted first in the cold Antarctic and Arctic areas because the cold temperatures create clouds of ice crystals in the stratosphere. Chlorine-containing compounds and ozone are brought together on the surface of these ice crystals and the third oxygen atom is removed to form oxygen gas. However, ozone depletion occurs to some degree worldwide, and preventing ozone depletion is a global effort.

The likely culprits of ozone depletion are hydrocarbon molecules that also contain chlorine or fluorine. They are particularly dangerous for two reasons: CFCs are very persistent and, because they have a catalytic role, are not used up in the ozone-depleting reaction.

Oddly enough, ozone is also a greenhouse gas. Without it, the stratosphere gets even colder, which in turn leads to a higher rate of ozone depletion. (See a more detailed description in Chapter 10.)

Weather and Pollution

The **grasshopper effect** is an environmental phenomenon and has nothing to do with grasshoppers. The grasshopper effect refers to the process by which certain chemicals are transported from warmer regions to colder regions of our planet. The grasshopper effect is driven by two physical principles: the convection cycles in the atmosphere and the variable solubility of toxins in water at different temperatures. The cumulative result of both these principles is that toxins tend to be taken into the atmosphere in warmer climates and released from the atmosphere in cooler climates. This results in a net transfer of pollutants from milder climates in lower latitudes to cooler climates at higher latitudes.

Inversions occur when temperature inversions trap pollutants close to the ground and a blanket of warm air prevents mixing and dispersal of cooler air underneath. There are two types of inversions: a *subsidence inversion* and a *radiation inversion*. Subsidence inversions occur over a broad area when a less dense warm front moves over the top of cool air and traps it, like a bubble. Radiation inversions occur over a smaller area when, as the sun sets, the air near the ground cools faster than the air further up. The warmer air traps pollutants in the cooler air underneath.

Heat domes occur in urban areas where asphalt and concrete-covered land absorb large amounts of heat that is re-radiated at night. This heat retention creates an island of heat around a city that deflects weather that would otherwise disperse pollutants. As a result, pollutants are held close to the city and are not as easily swept away by prevailing winds. (See a more detailed description in Chapter 10.)

Relevant Laws

1. **Clean Air Act (1970)** established air quality standards for primary and secondary pollutants; established "criteria pollutants" as the most threatening to human health; and requires states to develop clean air plans, which include the emission testing of cars.

2. **Montreal Protocol (1987)** is an international agreement that sets a timetable for phasing out the use of ozone-depleting substances.

3. **Kyoto Protocol (1997)** is an international agreement among 150 nations to reduce greenhouse gases.

> **TEST TIP**
>
> The laws and international agreements mentioned above are must-know items for the AP Environmental Science exam. They cover the major air pollutants and set the international goal of reducing globally destructive pollutants. You can be sure these laws will show up in some form on the AP Environmental Science exam.

Noise Pollution

Noise pollution is excessive, displeasing human, animal, or machine-created environmental noise that disrupts the activity or balance of human or animal life. Sources of noise pollution stem mainly from construction machinery and equipment and various types of transportation vehicles (primarily planes, cars, and trains).

The effects of noise pollution include hearing loss, elevated blood pressure, and loss of sleep in humans. When high noise levels exist in a community, it may have an impact on wildlife habitat and migration, thereby affecting biodiversity.

Control measures include installing highway barriers that insulate residential neighborhoods from car noise, low-noise jet engines that limit aircraft noise over populated areas, and workplace requirements for ear protection to protect individuals from occupational noise.

The **Noise Control Act (1972)** establishes standards for noise control and defines the federal government's role in research and enforcement of noise control.

Water Pollution

Types and Sources of Water Pollution

1. **Point sources** discharge pollution from specific points or locations. For example, drain pipes, sewer outfalls, and industrial drains are all considered point sources.

2. **Nonpoint sources** of water pollution come from sources that are not limited to a discrete, identifiable location. For example, rain runoff in a city washes sediment, oils, and heavy metals into local streams and rivers, but it does not come from any one source. Pesticides entering the groundwater come from many fields. It is difficult to identify a single source. Acid rain is caused by reactions that occur in the atmosphere, and then travels many miles before polluting a stream or river. It is a major nonpoint source of water pollution.

3. **Pathogens** represent a form of water pollution that poses the highest threat to human health. Most pathogens come from either a microbe contained in the life cycle of an insect or water organism, or from improperly treated human waste. Bacterial pathogens include those that cause cholera, dysentery, enteritis, and typhoid. Human coliform bacteria—such as *E. coli* and *giardia*, a protozoan—cause severe intestinal reactions, but they do not cause death. Viral pathogens include those that cause hepatitis A and polio. Schistosomiasis is one of the most prevalent aquatic pathogens in developing countries and is caused by an animal (blood fluke); it has increased drastically in countries that have built very large water projects, such as the Aswan Dam in Egypt. The slower-moving impounded water allows the growth of small snails, which are the vectors for the fluke.

4. **Oxygen-demanding wastes** come from the addition of nutrients, such as nitrates and phosphates or human waste, which help certain types of algae grow. Once the algae blooms, decomposing organisms break down the algal bodies and use up oxygen, which makes it difficult for organisms requiring respiration to live. Therefore, anything that encourages the growth of algae robs a waterway of essential oxygen.

    a. **Biological oxygen demand (BOD)** is a good estimate of the load of oxygen consuming organisms in a stream or river.

    b. **Dissolved oxygen (DO)** is a direct measure of the oxygen available for organisms in the water.

    c. **Oxygen sag curve in rivers and streams** occurs when oxygen-demanding waste is discarded into a river or stream, and the levels of DO drop downstream and create an *oxygen sag*. As the oxygen levels drop, so does the diversity and abundance of oxygen-demanding organisms. The *decomposition zone* exists immediately downstream from the discharge and contains organisms like leeches and trash fish, such as catfish.

The next zone is the *septic zone*, where the lowest DO exists. Fish cannot survive in the septic zone, and only highly tolerant organisms, such as worms and mosquito larvae, are hardy enough to exist.

As water continues to move downstream, the natural turbulence in the water helps replenish dissolved oxygen in the water. The colder the water, the more easily oxygen will dissolve. After the septic zone, the *recovery zone* represents the area where the DO returns to pre-discharge levels.

When the water has fully recovered, the *clean zone* is once again able to support sensitive, oxygen-demanding organisms, such as mayfly larvae, stonefly larvae, and cadis fly larvae.

Similar to lakes, rivers that have clear water, *low nutrient levels*, and low biological productivity are considered *oligotrophic*. *Eutrophic* rivers are nutrient-rich and carry a higher BOD. When eutrophication is accelerated as a result of human activity, it is considered *cultural eutrophication*.

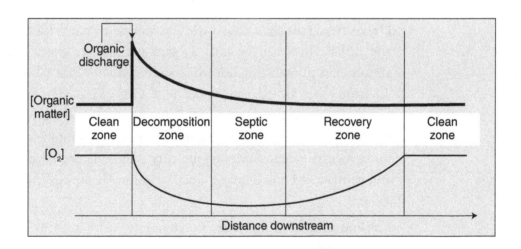

Figure 9-1. Oxygen Sag Curve in a River or Stream

5. **Inorganic wastes** are of two general types: suspended particles and dissolved ions, of which acid deposition is one type.

   a. **Suspended particles** are undissolved solids, such as small objects and sediment. Sediment that has washed into a stream or river from runoff or erosion is responsible for spoiling more waterways than any other type of water pollution. Sediment spoils the water for drinking and can cover or clog organisms that need a free flow of water or exposure to sunlight in order to survive. Turbidity is a measure of biotic and abiotic suspended solids; it is measured by determining the depth at which a black and white pattern on a secchi disk is still visible.

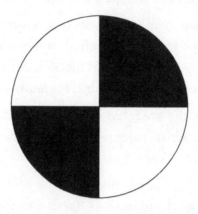

Figure 9-2. Secchi Disk Pattern

b. **Dissolved ions**

- **Heavy metal ions** are usually toxic and include mercury, lead, cadmium, and nickel. They can be fatal at parts-per-million levels. Metal ions already exist in their simplest form, so they do not break down further. Therefore, they are extremely persistent. One of the most famous cases of metal poisoning occurred in Minamata, Japan, where residents ate fish that had ingested mercury and then metabolized into the more toxic form, methyl-mercury. Methyl-mercury poisoning caused birth defects and permanent brain damage. Such poisoning is now called *Minamata disease*.

- **Calcium ions** spoil the water by combining with soap to decrease its cleaning ability and create a bothersome film. Additionally, calcium ions can also precipitate with other ions inside appliances to form a calciferous crust that may eventually destroy the appliance. For these reasons, many families use a "water softener" to replace the calcium ions with sodium ions so that appliances and soap work better. The drawback to this practice is that the sodium in the water can increase the incidence of hypertension in some people. Water containing calcium ions is usually called *hard water*.

- **Iron and sulfide ions** are found in areas where water is pumped from mineral-rich aquifers. Sulfide ions give water a rotten egg smell, but they can be removed with carbon filters or simple aeration. Iron ions affect the taste of water and can be harmful at high concentrations.

- **Chloride ions** enter the water as runoff from salted roads and industrial processes. If combined with organic material, such as particulate material from sewage treatment facilities, the resulting compounds may be carcinogenic.

- **Acid deposition** is perhaps the most damaging type of dissolved ion pollution. Acid deposition results from acid rain (reviewed earlier in this chapter) or from runoff that has filtered through solids that contain acidic material, such as mine tailings. Acid deposition threatens the stable pH that aquatic organisms need in order to survive. The aquatic larval stages of insects are more sensitive to fluctuating pH; severe damage to these ecologically important organisms can result in a drop of pH—or increase in acidity.

6. **Toxic organic wastes** can come from either natural or anthropogenic sources. One natural source of organic toxins is the so-called red tide, which is caused by a microorganism called dinoflagellate. Blooms of this organism are caused by nutrient wastes reaching marine pelagic ecosystems. This organism can attack fish and can also be toxic to humans who eat the fish. The resulting disease is called *paralytic shellfish poisoning*.

    Some artificially developed organic compounds represent the most toxic substances known. Water pollution as a result of these compounds presents a serious threat to human and nonhuman life. Pesticides that wash off agricultural and residential land are some of the most prevalent culprits. Organic wastes that leach out of dumps and corroded fuel tanks into the groundwater often end up in streams and rivers as well. Also, dioxins from burning trash can rain down on lakes and streams; exposure to these organic wastes at extremely low levels—a few parts per quadrillion—can lead to cancer and birth defects.

    Oil spills are another type of toxic organic waste. They are most often a problem in the open ocean or on navigable rivers that support barge and large craft transportation. Oil is a mixture of different hydrocarbons, most of which are carcinogenic to humans, are toxic to wildlife, and prevent photosynthesis by covering aquatic plants with a film.

7. **Thermal pollution**

    Cool water is able to dissolve more oxygen, which can then be used by aquatic organisms. As water heats up, less oxygen can dissolve in the water. When power plants use river water to cool their processors, it often passes through a turbine and the cooling water is returned to the river a few degrees warmer. As a result, the warmer water in the river decreases the river's DO levels, and something similar to an oxygen-sag curve results. In metropolitan areas, thermal pollution is most pronounced when it occurs alongside nutrient pollution, both of which decrease the DO of a river or stream.

Table 9-4. Summary of Major Water Pollutants

Pollutant	Examples	How Measured
Pathogens	Fecal coliform	Cultured water sample
Oxygen-demanding wastes	Nitrates, phosphates	Winkler titration
Inorganic wastes	Heavy metals	Chemical test
Dissolved ions	Ca^{+2}, H^+, S^{-2}, Fe^{+3}	pH for H^+
Suspended particles	Silt	Secchi disk, spectrophotometer
Toxic organic wastes	Some pesticides	Spectrophotometer
Thermal pollution	Heat from power plants	Thermometer

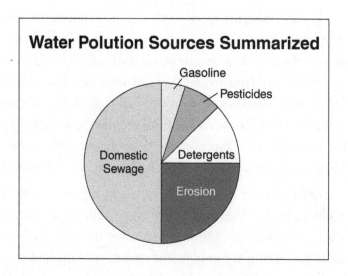

Groundwater Pollution

Any of the types of water pollution mentioned above can affect groundwater, or water that moves toward and through aquifers. The dynamics of groundwater pollution is often determined by the type of soil or rock the water flows through.

Groundwater pollution is particularly pernicious because it affects so many people. One cracked oil transport pipe, one fracking well, or one toxic spill can affect many people over a long period of time. Groundwater pollution creates a plume, just like air pollution that comes out of a smokestack. The plume's direction of travel must be taken into consideration if the plume is to be stopped. Specific methods of abating groundwater pollution are outlined later in this chapter.

Maintaining Water Quality

1. **Water purification** refers to the processes applied to naturally occurring water before humans use or drink it. Municipalities employ several different methods to purify water. Some of those methods are outlined below.
    a. **Filtration** of the water through course gravel and sand of various diameters removes particulate matter. Filtration through a carbon filter toward the end of the water purification process significantly improves the taste and quality of the water.
    b. **Flocculation** is the process by which aluminum sulfate is added to the water to bind to small particles so that they may be filtered more easily out of the water.
    c. **Disinfection** may occur before water enters the water supply. Chlorine, ozone, or UV light are typically used in this process.
2. **Sewage treatment** is composed of three stages of treatment: primary, secondary, and tertiary. Different municipalities use different combinations of these treatments in treating waste from human sewage. Some municipalities only use primary treatment methods to reduce costs or because they feel that they can get away with releasing partially treated sewage into a natural ecosystem.
    a. **Primary treatments** remove large particles through filtration and then allow bacteria to partially digest carbon and nitrogen wastes in large settling tanks. However, this does not entirely degrade the carbon into methane, or the ammonia/nitrogen into atmospheric nitrogen. Pathogens or toxins may also

still exist in the water. A primary treatment is fundamentally a filtration and settling process.

b. **Secondary treatments** hold the waste for a longer time in conditions that are favorable to bacterial digestion of the carbon and nitrogen wastes, thus lowering the BOD, or biological oxygen demand. The carbon-rich sludge tends to settle in the initial pond and can be removed to be digested by anaerobic methane-producing bacteria. The methane has a pungent odor, but it can be burned off or used to operate a generator to produce electricity. The nitrogen-rich solution is skimmed off and sprayed over high surface area substrate to promote aerobic, or oxygen-using, bacterial growth. These bacteria add oxygen to the ammonia form of nitrogen and convert it to nitrate ions. The carbon- and nitrogen-consuming bacteria can be easily filtered off before the waste is discharged. This solid waste is called *sludge* and presents a major disposal problem for many metropolitan areas. Also, the nitrates remain in the water and, if left untreated, lead to decreasing the available dissolved oxygen in the water after it is discharged. Pathogens can be killed by exposing the effluent to chlorine, UV light, or ozone before being discharged. Secondary treatment removes the greatest percentage of waste and pathogens, but it does not remove nutrients or toxic wastes.

c. **Tertiary treatments** use many different methods to remove nitrates, phosphates, and industrial pollutants. Nitrates and phosphates are fertilizers, and can be removed by sprinkling the water on trees or fields, or running the water through marshes so the plants assimilate the nutrients. Exposing the water to denitrifying bacteria converts nitrates into nitrogen gas. Passing the water between electrically charged plates removes the charged particles. Reverse osmosis can remove all dissolved and undissolved particles, but it is very energy intensive and expensive. A tertiary treatment may also involve specific chemical treatment to disinfect the water or remove toxic wastes, depending on the needs of the community.

3. Actions that can be taken to control water pollution include the following:
- Reduce emissions of sulfur and nitrogen oxides
- Modify agricultural practices
- Treat waste at industrial point sources
- Separate storm runoff from septic treatment
- Decrease silt runoff
- Reverse river channelization

Relevant Laws

The **Clean Water Act** was passed to "restore and maintain the chemical, physical, and biological integrity of the nation's waters." This legislation endeavors to make all waters "fishable and swimmable." It requires major polluters to have discharge permits it identifies toxic pollutants and requires best practices to remove them, and it sets goals for future development of new removal technologies.

The **Safe Drinking Water Act** maintains the quality of bodies of water—both on the surface and underground—that may be used for drinking.

Solid Waste

The **waste stream** is a steady flow of matter from raw materials, through manufacturing, product formation, and marketing, to the consumer and on to its final resting place—usually a solid waste dump. This is sometimes called a "cradle to grave" concern for waste management. At times, the consumer experiences a very short span of time in using the product on its way to the dump. For example, some junk mail may be in the hands of the consumer less than a minute before it becomes trash.

Americans process about 4.5 pounds of solid waste per day per person. About 76% of that waste ends up in a landfill. How we use our materials has a huge impact in the amount of waste we generate.

Types of Solid Waste

Solid waste is trash that is produced and disposed of through garbage collection. As the chart below indicates, paper is responsible, by far, for the greatest amount of solid waste. What is the second most voluminous type of solid waste?

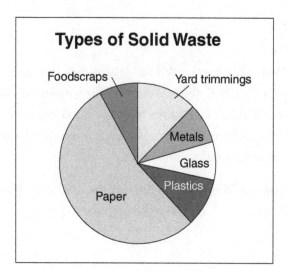

Waste Disposal

What are other options for getting rid of waste if it is not recycled?

1. **Dumps and Landfills.** While most developing countries use open, unlined dumps, which allow water to seep through the waste and contaminate the groundwater, developed countries usually use lined, sanitary landfills. At a well-run landfill, metal objects, batteries, burnable trash, rubber tires, and various

other recyclable or reusable wastes are sorted out. The leftover waste is then compacted and covered with dirt. The landfill is lined with clay or thick plastic. Wells drilled around the landfill allow monitoring of the groundwater nearby. Vents within the landfill allow methane gas—a by-product of bacterial anaerobic decomposition of carbon-containing waste—to escape. Some landfills even use methane to operate generators and generate electricity.

2. **Incineration.** Burning waste may be used by some cities to generate electricity or steam. This seems like a good idea. However, burning trash releases a class of toxins known as *dioxins*. Lead, mercury, furans, and cadmium have also been found in the ash of incinerated trash. The EPA approximates that toxic emissions of dioxin from a typical incinerator causes less than one death per million over 70 years—making it an "acceptable risk." Removing batteries decreases the risk due to the elimination of lead and other heavy metals, but it is impossible to remove all the plastics from the combustion, which is what would be needed to eliminate dioxin production.

3. **Toxic Colonialism: Sending Waste to Poor Countries.** As unethical as it may seem, several garbage producers have found that it is easiest to pay a poor country to accept its waste, or even just dump the waste without permission. In either case, taking advantage of an economically disadvantaged country in order to dispose of waste is called *toxic colonialism*. In some cases, where race correlates with lower socioeconomic power, disposing of waste in such an opportunistic manner —whether on foreign or domestic soil—is also called *environmental racism*.

Some countries are in such financial straits that they are willing to take cash in the short term in exchange for the long-term health of their people. For example, in 1986, the cargo ship *Khian Sea* left Philadelphia with a cargo of 28,000 pounds of incinerator ash. It was a last-ditch effort by the city because local states refused to accept the waste. The *Khian Sea* dumped a portion of the cargo in Haiti, but the Haitian government ceased the offloading when Greenpeace alerted them of the toxic nature of the cargo. After leaving Haiti, the *Khian Sea* visited Senegal, Morocco, Yugoslavia, and other ports in developing countries. For two years, the *Khian Sea* roamed the seas until, finally, it mysteriously lost its cargo—presumably dumped in the sea.

In 1999, 600,000 pounds of incinerator waste from Taiwan was dumped in Cambodia after the owners paid a $3 million "fee" to Cambodian officials. The real payment came when villagers near the dump site began dying of nerve and respiratory problems, and eventually had to be evacuated from their homes.

Reduce, Reuse, Recycle

Reduce, reuse, recycle is the sequence that leads to the least amount of overall use of energy and materials, and minimizes pollution.

> **DID YOU KNOW?**
> One print run of the Sunday edition of *The New York Times* requires the newsprint made from 63,000 trees. A single copy of the paper requires 280 gallons of fresh water to produce.

1. **Conservation and reduction** represent the most efficient ways to minimize the use of energy and minimize pollution. Simply consuming less has the largest impact. Here are some methods to reduce consumption levels:

   - Purchase materials with less packaging, or with no packaging at all, as in the case of food at a farmer's market.
   - Don't buy a new car very often; make them last. Cars use an extraordinary amount of plastic, metal, and chemicals—even water—to produce.
   - Ask cashiers not to give you a bag. Bring your own instead.
   - Use books from a library or online rather than purchasing new ones.
   - Check the news online rather than subscribe to a paper.
   - Don't make long trips for small reasons.

2. **Reuse** of products that have already been manufactured requires more energy and materials than not purchasing them at all, but far less than recycling. Here are a few practices that a typical consumer can reuse.

   - Reuse your own container instead of getting bags at the store.
   - Use a refillable water bottle rather than purchasing water in plastic bottles.
   - Use old clothes for dust mops.
   - Use aluminum foil more than once.

   > **DID YOU KNOW?**
   > When a household recycles glass, metal, plastic, and paper and also composts kitchen waste and yard clippings, very little garbage remains!

3. **Recycle**

   - **Solid waste** can be recycled in most metropolitan areas. This includes glass, metal, plastic, and paper. Recycling material saves raw materials and also the energy required to extract and transport those materials.

- **Composting** uses microorganisms and decomposing invertebrates to accelerate the decay of organic material. Materials that can be composted include grass clippings, leaves, farm animal manure, vegetable parts, coffee, egg shells, fireplace ash, and vacuum lint. Unacceptable materials include meats, oils, bones, plastic, glass, metal, pine needles, and pet waste. Decomposers use carbon and nitrogen from the waste and oxygen from the air to reduce the waste to simple nutrients, which can then reenter the nutrient cycles.

DIDYOUKNOW?

There are several variables involved in managing a healthy compost pile. The ideal carbon:nitrogen ratio is 30:1. If the compost smells like ammonia, then there is too much nitrogen from sources like grass and food scraps. If the compost is too rich with carbon, it will decompose very slowly. Oxygen is needed for the decomposers to undergo respiration and also helps to minimize odors. A healthy compost pile becomes quite warm, often over 100°F.

Effects of Pollution on the Environment and Humans

Precautionary Principle

The Precautionary Principle states that if there is some question about whether or not an action is hazardous to human or environmental health, it is the responsibility of the person who wishes to take action to prove that the action is safe. Another way to state the Precautionary Principle is to say that if there is uncertainty about whether or not an action is safe, then there is not enough knowledge to move ahead with the action. This is particularly true as communities face issues with unknown outcomes, such as whether or not to use hydraulic fracking to obtain more from oil and gas wells when others would experience damage from groundwater pollution.

Hazards to Human Health

Different hazards to human health include pathogens, toxic materials, and hazardous materials. The impact that they have on human health is measured by morbidity and mortality. There are a number of ways to assess the risk of morbidity and mortality that environmental hazards may pose.

1. **Morbidity and mortality:** Altering human health causes disease, or *morbidity*. If disease is severe enough, it causes death, or *mortality*. Morbidity and mortality are measured in the number of occurrences per thousand per year. If the mortality of a population is 16, that means that 16 people per thousand in the total population die each year.

2. **Environmental risk analysis** is important in determining causes of morbidity and mortality. These types of experiments can help identify the source, magnitude, and mechanism of action of pathogens and toxins that cause disease.

   - **Bioassay** is a biological test for toxicity. Such a test may look at toxicity of a toxin on a number of species, whether they are plant, animal, fungus, or bacteria. For example, constructing a dose-response curve (see below) using mice is one type of bioassay. Another would be to test antibiotics on colonies of different bacteria.

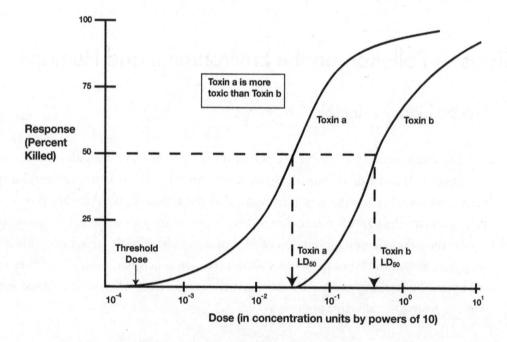

Figure 9-3. Typical sigmoidal dose-response curve. The horizontal axis is usually a logarithmic scale. The vertical axis represents the response—which could refer to anything from being healed from a disease to contracting cancer. When the response recorded on the vertical axis is death, the dose that kills 50% of the subjects is called the lethal dose 50%, or LD_{50}.

- **Disease clusters** can be used ethically to determine the cause of disease on a human population. While it would not be ethical to give humans a disease on purpose, observing how diseases emerge in clusters certainly gives researches a clue about what may cause a disease. A scatterplot may be used to show correlation between disease in the neighborhood surrounding a source of pollution. Once a disease cluster is identified and some statistical correlation with environmental factors is established, further study may reveal an actual cause-effect relationship between the disease in a particular individual and a specific pollutant.

- **Dose-response curves** quantify the toxicity with particular bioassay experiments. The concentration of the toxin represents the independent variable on the horizontal axis, and the response of the organism is catalogued on the vertical axis.

- **LD_{50}** (lethal dose for 50% of the sample) is found on the dose-response curve when the "response" is the death of the organism. The LD_{50} is that dose that corresponds to the death of 50% of the sample size. One of the drawbacks of using LD_{50} to identify toxicity is that it requires the death of a large number of organisms, which may present an ethical issue, depending on the species. Also LD_{50} may show toxicity in one type of test organism, but the physiology of the test organism may be quite different than for humans. Also, while LD_{50} gives a crude measure of toxicity, it does not suggest how an organism is toxic. Is it a teratogen? Carcinogen? Mutagen? The LD_{50} gives no hint about the *mechanism* of toxicity.

- **Threshold level** of toxicity is determined from a dose-response curve, which represents the dose at which none of the test subjects are harmed. With respect to drugs, finding this level in mammals helps to guide researches for setting human tolerance levels.

> **DIDYOUKNOW?**
> Over 100 years ago, miners would carry a canary with them into a mine. Canaries are very sensitive to carbon monoxide poisoning, which is a constant concern — particularly in a coal mine. If a miner's canary passed out, he knew it was time for him to leave the mine. Even though coal miners no longer use canaries, the concept of using animals to predict health hazards still continues.

TEST TIP

Understanding how to obtain the LD_{50} is a key concept that frequently shows up on AP Environmental Science exams. Be sure that you can read a dose-response curve and can determine which of two graphs represents the most toxic compound.

3. **Pathogens** are disease-causing organisms or viruses. Every living kingdom contains some members that can potentially cause disease. The following are important aspects of pathogens that relate to environmental science.

    - **Vectors** are organisms that have a symbiotic relationship with a pathogen and therefore can carry the pathogen from one host to another. Quite often efforts to eradicate disease focus on eradicating the vector, as was the case for malaria mosquitoes, the vectors for the malaria-causing microorganism.

    - **Vaccines** help prevent disease by introducing a foreign *antigen* into the system that stimulates the production of human *antibodies*, which are then available should an actual infection take place.

    - **Resistance** can develop among pathogens when mutant strains develop that are no longer affected by pesticides or antibiotics. For a time, malaria was almost entirely eradicated, but it now affects over 3 million people per year because the vector mosquitoes have become resistant to the pesticides.

    - **Emergent diseases** are new to a population, or have been absent for some time and have returned. Emergent diseases can pose a large threat to human populations because of the rapid spread in today's rapid-transit global society. In addition, there is always a chance that a new disease may develop for which we have no latent protection within our genes.

    - **Disease virulence** refers to the strength of a pathogen, or its ability to cause death. Even the most virulent diseases decrease in severity over time. Lethal versions of pathogens have a harder time surviving after a host dies, so less lethal versions tend to spread more and thus decrease the virulence in each subsequent generation.

4. **Toxic materials** may have one of the following types of mechanisms that cause disease.

    - **Mutagens and carcinogens** can both cause cancer. *Mutagens* cause cancer by changing the DNA in living cells. *Carcinogens* cause cancer by either turning ON a cancer-causing gene, called an *oncogene*, or turning OFF a cancer-suppressing gene, called a *tumor suppressor gene*. In either case, a single cell needs multiple mutations to cause cancer. These mutations may occur from environmental toxins. Ultraviolet light, nicotine, radiation, and some hydrocarbons are examples of carcinogens.

- **Neurotoxins** attack nerve cells, which may be in the central nervous systems (brain and spinal cord) or in the peripheral nervous system. Because the nervous system is in control of every aspect of human physiology, nerve toxins can be very damaging and fast acting. Nerve cells can be affected by toxins in many ways. Heavy metals block receptors so that neurotransmitters cannot bind to them, for example. Some toxins, such as a category called anticholinesterases, which are present in many pesticides, prevent neurotransmitters from being recycled and eventually close down the nerve impulse.

- **Allergens** are recognized by the immune system as a foreign object. Then the immune system responds to try to get rid of it. One of the chemicals secreted by the immune system is *histamine*, which causes cold-like symptoms. Sometimes these symptoms are worse than the foreign particle, so people take an *antihistamine* to minimize the symptoms.

- **Teratogens** cause birth defects. Perhaps the most infamous incident of teratogen exposure occurred in the 1960s, when thalidomide was sold as a sleeping aid. Unfortunately, even a small amount of this drug in the early weeks of pregnancy prevented limb development, and children were born with a small hand or foot, or no arm or leg. Alcohol during pregnancy can cause fetal alcohol syndrome, which is demonstrated by delayed development and mental disabilities.

5. **Factors that affect toxicity:**
    - **Dose** refers to the amount of a drug or toxin ingested multiplied by the time span during which exposure occurs. The larger the dose, the larger the effect.
    - **Genetic predisposition**, or an individual's genetic makeup, strength of immune and excretory systems, and overall hardiness due to nutrition and general health all affect how severely a toxin is experienced.
    - **Chemical synergy** occurs when two toxins together have a greater effect than the sum of the effects of the two toxins separately. For example, a moderate amount of alcohol can become deadly when combined with even a low level of barbiturates. In another example, while smoking and asbestos can both cause lung cancer, being exposed to both increases the risk of cancer by a factor of ten!

- **Solubility** is an important physical property that determines whether or not a toxin remains in an individual's body and/or is passed on to the next trophic level. If a toxin is water-soluble, it is readily excreted and does not stay with living tissues. However, it is more mobile in ecosystems. If a toxin is fat-soluble, it is retained in living tissue but is less mobile in ecosystems. When fat-soluble toxins are retained in living tissue, they *bioaccumulate* in the organism. Bioaccumulation increases toxicity for that organism because the toxin exerts its toxicity over a longer period of time, and more toxin is likely to be retained—both of which increase the overall dose of the toxin. Furthermore, fat-soluble toxins that bioaccumulate within a single trophic level can also *biomagnify* as they pass to the next trophic level. *Biomagnification* most significantly affects top carnivores, even if the substance isn't toxic to the producer or primary consumer.

- **Persistence** refers to the toxin's ability to remain active as it moves through the environment. If the toxin degrades into a less toxic version from bacterial digestion or exposure to sunshine, it is not as persistent. For example, DDT was a persistent pesticide so it did not need to be applied frequently. However, its persistence and fat-solubility allowed it to remain in ecosystems in its toxic form and to biomagnify as it passed from one trophic level to the next.

- **Acute and chronic effects.** A toxin dose that inflicts immediate harm to an organism demonstrates *acute toxicity*. A smaller dose over a long period of time may cause *chronic toxicity*. It is easier to design a study that establishes acute toxicity because the cause for the poisoning is more easily ascribed to the toxin. Chronic toxicity is harder to detect because the effects might not be seen for years, and it is difficult to isolate all exposures over those years to guarantee which exposure caused illness.

Table 9-5. Summary of Factors That Affect Toxicity

Factor	Explanation
Dose	Equals amount over length of time administered.
Genetics	Some people are more susceptible.
Synergy	Combinations of toxins can sometimes be more toxic than each toxin separately.
Solubility	Fat-soluble toxins bioaccumulate, increasing the dose over time.
Persistence	Some toxins are slower to break down when in the body or the environment.

Hazardous Chemicals in the Environment

1. **Types of hazardous waste** are distinguished from toxic wastes because they pose an explosion, burning, or oxidizing hazard—rather than a disease-causing toxicity.

    a. **Ignitability** refers to the ability of a material to combust. Examples include gasoline, kerosene, or paint thinner. Sometimes only the fumes of the compounds are needed to start a fire. The hazard stems from the risk of burns and smoke inhalation.

    b. **Corrosiveness** refers to the ability of a material to act as an oxidizing agent. This action corrodes metals and burns skin. This is the case with strong acids and chlorinated compounds. When disposing of these compounds, they must be isolated from other compounds and put into containers that do not corrode or leak.

    c. **Reactivity** refers to materials that easily undergo a rapid chemical reaction and explode. Compounds in this class include gunpowder, carbides, sodium metal, nitroglycerine, and peroxides. They must be protected from the trigger that causes them to explode. For example, sodium metal must not get moist because water causes it to oxidize rapidly, releasing heat and hydrogen gas—which then explodes from the heat.

2. **Natural remediation of hazardous and toxic waste:**

    a. **Treatment in the atmosphere** can occur in one of three ways: *photolysis, oxidation*, or *precipitation*. Some toxins are broken down by high-energy ultraviolet rays from the sun, or *photolysis*. Solar rays may combine with the oxygen in the atmosphere to *oxidize* the toxin, reacting to form a less hazardous compound. Sometimes toxins are simply washed out of the atmosphere by rain. For example, pollen is a potent allergen that is cleansed during rain.

    b. **Treatment in the water** can be done by hydrolysis or microbial digestion, among other ways. Some toxins actually degrade by reacting with water in a hydrolysis reaction. Some toxins become food for bacteria, which is then secreted in a less toxic form. Bacterial degradation is one of the most effective ways to remediate environmental accidents, such as oil spills. Sometimes toxic wastes simply sink to the bottom and become covered with sediment. For example, some waterways near industrial areas become more polluted after the area has been dredged—a process that removes the mud and kicks up buried toxins.

c. **Treatment in the soil** is often accomplished by bacterial degradation as toxins percolate through the soil. For this reason, water in deep aquifers tends to be very clean. By the time water has percolated through hundreds of feet of soil and rock, bacteria have been able to digest and metabolize the toxins.

3. **Cleanup of contaminated sites**

   With the passage of CERCLA in 1980 (see below), funds became available to clean up massively contaminated sites across the United States. There are several ways to remediate a toxic, contaminated site. Some are summarized below. In whatever way remediation occurs, it is an extensive and expensive operation.

   a. **Reduce, reuse, recycle** can also be applied to hazardous waste. The less that is produced, the less needs remediation. If it can be reused in any way, then less waste will be produced. However, what can be done when—eventually—the waste must be cleaned? Perhaps the most effective method is *bioremediation*.

   b. **Bacterial remediation** uses bacteria to digest toxins in the ground. This process is sometimes also called *bioremediation*. Either naturally occurring bacteria, or bacteria developed artificially through genetic recombination, are used onsite or offsite to reduce the toxins in soil to less toxic substances.

   c. **Air stripping** is a process that presses highly pressurized warm air against the contaminated material, which has the effect of vaporizing toxic volatile organic compounds (VOCs) from the contaminated material and drawing it into the gas phase. This can be done offsite or onsite.

   d. **Capping** is simply burying the toxic portion of the site, and then "capping" it with a material—such as cement, clay, or thick plastic—that isolates the material from areas that will be exposed to people.

   e. **Incineration** is used to degrade toxic materials into nontoxic compounds. The contaminated substance is heated to a high temperature and the toxic compounds are combusted. The material is then returned to the site.

   f. **Precipitation** first dissolves toxic soluble ions in water, then passes that water through a solution that contains ions that combine with the toxic ions and form a solid precipitate. Then the precipitate is filtered out. This method can be used for heavy metal ions that form solid precipitates with carbonates, sulfates, and sulfides.

Relevant Laws

— **Federal Hazardous Substances Act (1960)** requires cautionary labels on household products.

— **Hazardous Materials Transportation Act (1975)** regulates the transport of hazardous materials.

— **Comprehensive Environmental Response, Compensation, and Liability Act (CERCLA) of 1980** establishes the "superfund," which is used to clean up toxic waste sites.

LEED Certification

How the Economics of Pollution and Waste Control Can Blend with Sustainable Building Practices

In other chapters, you have reviewed how external costs are significant in environmental issues and mechanisms by which external costs become internal costs. In this section, you will consider an example of how external costs are incorporated into sustainable building practices through the LEED certification program.

TEST TIP

While not a stated AP Environmental Science topic, it is clear that LEED certification of buildings is having a remarkable impact on building technology with respect to the amount of waste and pollution created and the amount of energy used by a building. Therefore, this section is included so that you can use LEED certification of buildings as an example in free-response questions.

What Is LEED Certification?

LEED is an acronym that stands for "Leadership in Energy and Environmental Design." LEED certification of buildings and professionals who work with these buildings was created by the United States Green Building Council (USGBC) to provide a credible group of standards for environmentally sensitive building design.

LEED certification is a voluntary process that leads to certification of a building at various levels of sustainability: Silver, gold, and platinum levels all represent increasingly sophisticated levels of sustainability.

LEED certification was designed to create buildings that are environmentally sustainable over a period of time. In the summary of standards below, there may be some that seem out of place (like having showers in the workplace, for example), but these standards were designed to reduce the overall use of materials, toxins, and energy over time (encouraging people to bike to work and reduce car emissions, for example). In short, LEED certification accomplishes the following:

1. Reduces waste sent to landfills
2. Conserves energy and water
3. Provides a healthier and safer environment for occupants
4. Reduces harmful greenhouse gas emissions

For a building to be LEED certified, a LEED-certified professional must be on the team and document each LEED requirement that has been met in the project. The requirements are summarized in the section below.

LEED Certification Requirements

LEED certification is attainable at different levels, based on points earned from the following checklist. (This checklist is for new construction and major renovations. There are similar, but slightly different, checklists for other types of projects.) To be certified, the building needs 10 to 49 points from this checklist. To achieve silver certification, 50 to 59 points are needed. To have gold certification, 60 to 79 points are required. Platinum certification requires over 80 points. (The following chart is adapted from the USGBC checklist to help summarize the important points for this review.)

Table 9-2. LEED Certification Points

Sustainable Sites	26 points
Site pollution	1
Community connectivity	5
Brownfield development	1
Transportation — public transport	6
Transportation — enable bicycle	1
Transportation — low emission cars	3
Transportation — parking	2
Protect/restore habitat	1
Increase open space	1
Stormwater control	2
Reduce heat island effect	2
Reduce light pollution	1
Water Efficiency	**10 points**
Use reduction	Required
Water efficient landscaping	2–4
Further reduction, innovation	4–6
Energy and Atmosphere	**35 points**
Optimize energy performance	1–19
Onsite renewable energy	1–7
Green power	2
Materials and Resources	**14 points**
Building reuse — keep existing walls, etc.	1–3
Building reuse — keep 50% interior	1
Construction waste management	1–2
Materials reuse	1–2
Recycled materials	1–2
Materials from nearby region	1–2
Rapidly renewable materials	1
Wood certified from sustainable forests	1

(continued)

(continued)

Indoor Environmental Quality	15 points
Outdoor air delivery monitoring	1
Increased ventilation	1
Manage air quality during construction	1
Manage air quality before occupation	1
Low-emitting adhesives, sealants	1
Low-emitting paints, coatings	1
Low-emitting flooring	1
Low-emitting wood, fiber products	1
Indoor pollution control	1
Lighting control	1
Heating/cooling control	1
Design of thermal comfort	1–2
Daylight and views	1–2
Innovation and Design Process	**6**
Compatible with Regional Priorities	**4**
TOTAL POSSIBLE POINTS	**110**

TEST TIP

While LEED certification is not on the AP Environmental Science syllabus, it is a constructive and efficient study tool to go through the LEED checklist above and consider how each point helps to reduce pollution, energy use, and/or waste.

Time for a quiz
- Review strategies in Chapter 2
- Take Quiz 6 at the REA Study Center

(www.rea.com/studycenter)

Chapter 10

Global Changes

The major global changes discussed in this chapter draw from the content of the previous six chapters. The topics involve information, usually from more than one chapter, that is applied to a far-reaching global phenomenon. Of the three major topics in this chapter, all will be on the AP Environmental Science exam in some manner, and it is highly likely that one of these three will be represented in some way among the free-response questions.

The College Board syllabus indicates that the following topics from this chapter may show up on the AP Environmental Science exam. These topics compose about 10% to 15% of the total test.

A. **Stratospheric Ozone**
   (Formation of stratospheric ozone; ultraviolet radiation; causes of ozone depletion; effects of ozone depletion; strategies for reducing ozone depletion; relevant laws and treaties)

B. **Global Warming**
   (Greenhouse gases and the greenhouse effect; impacts and consequences of global warming; reducing climate change; relevant laws and treaties)

C. **Biodiversity**
   1. Habitat loss; overuse; pollution; introduced species; endangered and extinct species
   2. Maintenance through conservation
   3. Relevant laws and treaties

Stratospheric Ozone

Formation of Stratospheric Ozone

For millions of years, blue-green algae underwent photosynthesis to form the molecules of oxygen gas (O_2) in the early atmosphere. These oxygen molecules were bombarded by ultraviolet (UV) light from the sun, splitting the oxygen molecule in a process called photolysis, and one of the atoms of oxygen was added to another molecule of O_2 to form ozone, or O_3.

Ozone (O_3) built up in the early atmosphere to eventually form enough of a barrier to dangerous UV light (in the 200-300 nm range) so that life could move out of the oceans—protected from the sun by water—and evolve on land. Scientists think that this level of protection probably developed about 600 million years ago. Now, the ozone layer provides a shield for all living organisms from UV light that would otherwise cause burns, cancer, and genetic damage.

Ultraviolet Radiation

Ultraviolet radiation represents a portion of the electromagnetic spectrum of light energy that comes to the Earth from the sun. The most damaging portion of the spectrum is the portion with a wavelength from 280 to 320 nm, or the UVB region, as shown in the following figure.

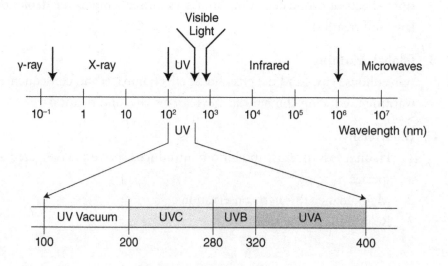

Figure 10-1. Spectrum of Light

Causes and Effects of Ozone Depletion

First reported in 1985, the periodic creation of a hole in the ozone layer in the stratosphere was startling news. Ozone tends to be depleted in the polar regions because the cold temperatures create clouds of ice crystals in the stratosphere. Chlorine-containing compounds and ozone are brought together on the surface of these ice crystals and the third oxygen atom is removed to form oxygen gas. However, ozone depletion occurs to some degree worldwide—not just in the polar regions—and preventing ozone depletion is now a global effort. The figure below illustrates how O_3, ozone, is converted to O_2, oxygen gas, in a reaction that is catalyzed by chlorofluorocarbons (CFCs).

To some degree, ozone depletion also occurs over summer storm clouds in the United States. As these storm clouds increase in size and frequency (due to increased water vapor in the atmosphere from global warming), ozone depletion from these clouds may increase.

At this time, scientists believe that the Earth is undergoing a depletion of global ozone equivalent to about 4% per decade, with much larger amounts around the polar regions during the spring. Depletion levels have begun to subside because of efforts to reduce the production and use of CFC molecules; however, it will be some time before the Earth reaches pre-1980 levels.

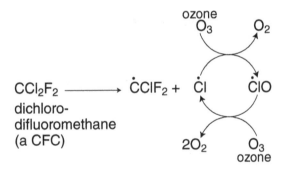

Figure 10-2. Conversion of O_3 to O_2

If the ozone layer did not exist to protect life, high-energy UV radiation would bombard our cells. The damage to our skin cells would cause sunburn at low doses, and skin cancer and immune system suppression at high doses. Also, UVB rays may also decrease phytoplankton production in the ocean. While the former effect certainly affects human individuals (particularly if they don't use sunscreen), the latter effect has broad-ranging global ecological consequences. Increased UVB light hitting the planet can also have an effect on terrestrial organisms such as those that depend on cyanobacteria (like rice), which is sensitive to UVB light.

Strategies for Reducing Ozone Depletion

Strategies for reducing ozone depletion focus on eliminating production and use of chlorinated and brominated fluorocarbons. These strategies have been so successful that the remaining molecule that continues to deplete ozone is nitrous oxide (N_2O), which is not covered by the Montreal Protocol (see below). New strategies need to be developed to reduce the use of nitrous oxide.

Relevant Laws and Treaties

The Montreal Protocol: The countries that participated in the Montreal Protocol agreed to freeze production of CFCs at 1986 levels, and reduce production of CFCs by 50% before 1999. At a later meeting, this agreement was strengthened to not just reduce—but cease—production and use of all CFCs, except for the most critical uses.

> **TEST TIP**
>
> The Montreal Protocol is a key international agreement to know and remember for the AP Environmental Science exam.

Global Warming

The Greenhouse Effect

When solar radiation penetrates the Earth's atmosphere, it is either reflected or absorbed. The reflected radiation—about one-third of the incoming solar energy—returns to space. The absorbed radiation causes thermal motion in matter on Earth (conduction). When matter vibrates, it reemits low-energy, long-wavelength electromagnetic radiation. This long-wavelength radiation returns to space under normal conditions. However, greenhouse gases in the atmosphere absorb the reemitted low-energy radiation as heat. The fact that this has happened has kept the temperature of the Earth warm and has actually helped life evolve. However, further production of greenhouse gases from industries is now causing more heat to be retained, and Earth's climate is changing.

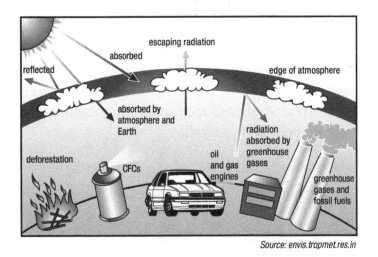

Figure 10-3. The Greenhouse Effect

Greenhouse gases include carbon dioxide, methane, water vapor, ozone, and chlorofluorocarbons. Carbon dioxide is particularly significant in climate change and the greenhouse effect. Carbon dioxide production has increased from our combustion of fossil fuels to produce heat and electricity. Using electricity that originates from a coal-fired power plant, driving to and from work in gas-thirsty cars, and using products and packaging that require energy to produce are all practices that increase the amount of carbon dioxide in the atmosphere.

Mechanisms That Accelerate Global Warming

In addition to humans putting more greenhouse gases into the atmosphere, there are other occurrences that accelerate global warming through a positive feedback loop. The more each of the following events occur, the greater the speed at which each of the other events occurs, and the rate of global warming accelerates.

1. **Methane sequestration**

    As polar regions warm, methane gas, which has been trapped for centuries in frozen tundra bogs, is released. In addition, methane hydrate is found within the lattice of ancient polar ice, which also releases methane when that ice melts. These two mechanisms release methane—a greenhouse gas—into the atmosphere and, as the polar regions warm, the warming of the area is accelerated.

2. **The Albedo effect**

    "Albedo" refers to the reflective nature of any surface. The more solar radiation ice reflects, the higher its albedo. Ice has a particularly high albedo. Polar ice reflects the sun's energy back into space, thereby preventing solar radiation from significantly heating that region. However, as polar ice melts, there is less albedo, less reflection of solar radiation, and greater absorption of heat, thus accelerating global warming as the ice melts.

3. **Deforestation**

    Carbon sequestration occurs as forests gulp huge amounts of carbon dioxide during photosynthesis and store that carbon in the form of biomass. As forests are being cut down, the decreased sequestration of carbon accelerates the buildup of atmospheric carbon dioxide, thus accelerating global warming.

4. **Water vapor production from increased evaporation**

    With increased surface sea temperatures, evaporation of water accelerates. Because water vapor is yet another greenhouse gas, the greater amount of water evaporation also accelerates the greenhouse effect—and global warming.

Evidence of Global Warming

While some people are not convinced that global warming is occurring, scientific data demonstrate that there is both a measurable increase in global air and sea surface temperatures, as well as an increase in measurable carbon dioxide—the major greenhouse gas of concern, and the gas that humans are most responsible for due to combustion of fuels and deforestation. Frequently, some combinations of the following types of data are used as examples on AP Environmental Science exams because together they form a compelling case for the idea that humans are causing the Earth's climate to change.

1. **Historic temperature data**

    The following data represents measurable changes in historic mean global temperatures since 1880.

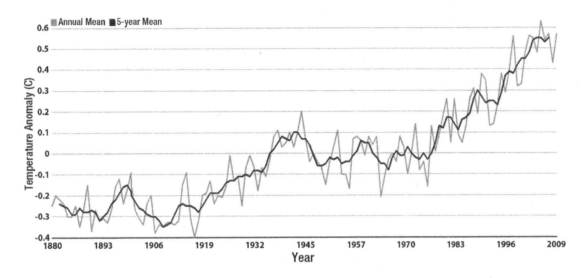

Figure 10-4. Historical Global Land-Ocean Temperatures, 1880 to 2009

2. **Measured increases in carbon dioxide**

    The data below are frequently cited on AP Environmental Science exams and come from carbon dioxide measurements taken from the Mauna Loa Observatory in Hawaii. Students are typically asked why the actual value fluctuates on an annual cycle. The most important point about this data comes at the 10 ppm increase in carbon dioxide, which is a major anthropogenic greenhouse gas. Note that the average annual value, as well as a monthly mean, is tabulated. The annual cycles are due to seasonal changes in photosynthesis occurring in the world's forests. (Remember, an increase in photosynthesis results in less atmospheric carbon dioxide.)

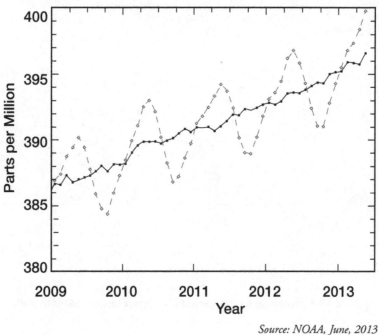

Figure 10-5. Recent Monthly Mean CO_2 at Mauna Loa

3. **Ice core data**

   Using the amount of carbon dioxide trapped as bubbles in polar ice, scientists are able to determine levels of carbon dioxide for several hundred years. There is a clear correlation between carbon dioxide levels and air temperatures over the last thousand years (see figure below). Toward the right, we see a steep increase in carbon dioxide levels that corresponds to the industrial revolution. This steep increase has a much larger magnitude than normal fluctuations of carbon dioxide and temperature over several millennia.

   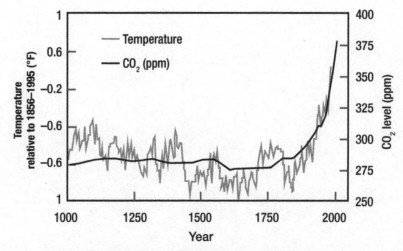

   Figure 10-6. Temperature and CO_2 for Last 1,000 Years

4. **Polar ice melting**

   There has been a significant change in polar ice in just a few years. The following figure shows how arctic ice has reduced just since 1978. The change in Antarctic sea ice is also dramatic, as is the change in Greenland land-based ice.

   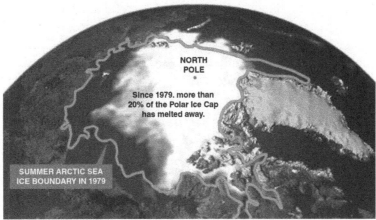

   Source: NASA

   Figure 10-7. Reduction of Arctic Ice since 1978

5. **Measurable sea level increases**

   A measurable increase in the sea level corresponds to decreased polar ice and an increase in sea surface temperatures, both of which add to the volume of the ocean. Remember that warmer water is less dense and occupies a larger volume.

6. **Measurable increases in sea surface temperatures**

   Increasing sea surface temperatures seem to indicate that global temperatures are indeed rising because the sea is a thermal reservoir of sorts. The increased sea surface temperatures also account for a greater number of hurricanes and other storms that are based upon the amount of atmospheric water vapor. The higher the sea surface temperature, the greater the amount of evaporated water that creates water vapor.

DIDYOUKNOW?

Some South Pacific Island countries, whose total elevation above sea level is just a few feet, are purchasing land in other countries so that the small cultures can continue to exist in the event of drastic rises in sea level.

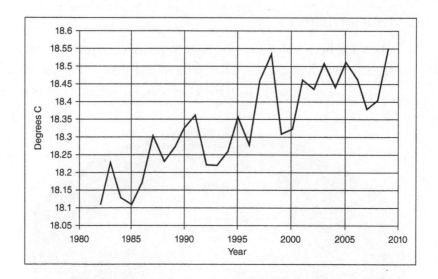

Figure 10-8. Global July Sea Surface Temperatures, 1982 to 2009

7. **Decreased ocean pH**

    Due to Henry's Law, the concentration of a dissolved molecule in a liquid is directly proportional to the partial pressure of the same molecule in the air above the liquid. Henry's Law suggests that increased amounts of carbon dioxide in water also means that there is more carbon dioxide dissolved in the ocean water. Historical data shows that this is, in fact, the case.

    An increase in carbon dioxide in the water results in an increase in dissolved H^+ ions in the ocean, or a more acidic (lower) pH. This is the result of adding CO_2 to the following carbonate buffer system that takes place in ocean water.

    $$CO_2 + H_2O \leftrightarrow HCO_3^- + H^+ \leftrightarrow CO_3^{-2} + H^+$$

    Increasing the amount of carbon dioxide to the left side of this set of chemical equations results in producing more H^+ ions.

8. **What about natural climate cycles?**

    Some are skeptical that global warming takes place, stating that the current measurable warming trend is one of many that have occurred as a result of large-scale global climate fluctuations. These fluctuations over many thousands of years have been measured using polar ice cores. These fluctuations are in the 0.02 ppm range. However, the current increase in carbon dioxide is in the 2.0 ppm range—about 100 times larger.

TEST TIP

Regardless of the politics of whether or not someone may believe the scientific evidence about climate change, the AP Environmental Science exam will use data from the current scientific community to ask questions about this topic. Those data will likely come from one of the above types of data sets. Students should be prepared to understand and use this data, and also consider some of the collateral results of climate change that are mentioned below.

Effects of Climate Change

1. **Alpine glaciers** melt or form as a result of long-term climate change. Currently, glaciers are retreating, which decreases freshwater runoff to local human populations.

2. **Arctic sea ice** melts or forms as a result of long-term climate change. Arctic sea ice is currently disappearing and releases methane gas bound up in the ancient ice, further increasing climate change. Additionally, the decrease in large amounts of reflective ice in the polar regions is reducing the amount of solar radiation that is being reflected, and more energy is being absorbed—further contributing to warming.

3. **Polar land-based ice** melting contributes to raising sea levels. Scientists are currently observing an acceleration of moving land-based ice because there is considerable water flowing underneath the ice, rather than the ice resting on solid ground. As the amount of melted ice underneath the solid ice increases, so do the chances that the ice will move very quickly into the sea, floating atop the river of water underneath it.

4. **Change in ocean currents.** The El Niño effect is just one global event that results in a shift of ocean currents. Scientists now feel that a rapid movement of Greenland ice into the North Atlantic may cause a major shift in the Gulf Stream current. Shifting ocean currents change climate conditions for the continents near them and shift the ecosystems around those regions.

5. **Ocean levels** change as a result of long-term climate change. In the current trend of melting alpine ice, ocean levels are rising. A 45-cm rise in ocean level would flood much of Florida, coastal areas in Bangladesh and Pakistan, and several small Pacific island countries. Major cities, such as New York, London, Calcutta,

Jakarta, and Manila would need to abandon expensive ground or undergo massive construction projects to protect themselves from the rising sea.

6. **Disease-carrying vectors** change their normal ranges as climate changes. West Nile virus is carried into areas that did not previously see this mosquito-borne disease, for example.

7. **Biodiversity is threatened** because each living species has evolved within certain conditions, and those conditions change with climate change. This is currently being observed on a number of levels all over the world. The following are a few example:

    a. **Caribou and mosquitoes**

    Caribou are the largest herding animal in the arctic and subarctic north. They use the summer months to feed on small tundra shrubs and can forage for some plants during the winter because it is too cold to snow. During the summer, the caribou fatten up for a few weeks before the mosquitoes get so bad that the caribou are driven to the high country for relief. Warmer winters bring greater snow depth and earlier summers bring more mosquitoes earlier in the summer. Between more snow and earlier mosquitoes, caribou are not consuming enough food to survive. Caribou herds decreased by 95% between 1960 and 1997, and much of this decline is likely from decreased food availability.

    b. **Specialist species** with narrow temperature tolerance ranges—particularly at high altitudes—will not be able to migrate quickly enough and will likely die out.

    c. **Coral reef bleaching** occurs as higher mean ocean temperatures shift and the pH of the water decreases. The higher sea temperatures increase coral bleaching and disease. The pH drops because the increased carbon dioxide in the atmosphere increases the amount of dissolved carbon dioxide in the water, and shifts the equilibrium between dissolved carbon dioxide and carbonic acid toward the production of more acid. The more acidified ocean also decreases the survivability of coral.

    The productivity of coral reefs is extremely high, and they support a high level of biodiversity. As reefs are threatened, so are the diverse organisms that depend on them.

8. **Permafrost melts** in tundra biomes when climate change increases average temperatures. This also releases methane into the atmosphere, which further contributes to climate change.

9. **Change in the frequency and strength of cyclonic storms and global precipitation** occurs as climate change increases surface sea temperatures, causing more water to evaporate. This may affect flooding in some regions and cause greater storm damage—particularly in hurricane prone regions of the world.

10. **Loss of coastline.** With great concentrations of the world's population living near coasts, rising coastal waters will have a significant effect on the densely populated coastal areas. This may result in a loss of property, industry, coastal ecosystems, and economic vitality.

Carbon Footprint

The carbon footprint is the total amount of greenhouse gases that are caused by an individual, organization, event, or product. Carbon footprint may be a difficult value to measure exactly; however, the concept is educational—particularly with respect to the indirect pollution inherent within the choices that people make. Which activities represent the top three contributors to your carbon footprint?

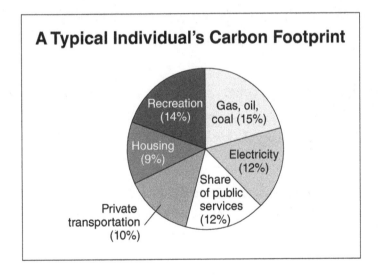

Reducing Climate Change

Carbon footprint—and therefore climate change—can be reduced by actions that reduce the use of greenhouse gases or decrease the decline of carbon sequestration.

1. Reduce production of greenhouse gases
   - Use less coal to provide electricity; instead, obtain electricity from renewable sources, such as solar and wind power.
   - Drive high-mileage cars, and therefore use less gasoline, which is a derivative of crude oil.
   - Use less plastic.
   - Build LEED-certified buildings.
   - Use public transportation.
   - Buy local; go on local vacations.

2. Reduce decline of carbon sequestration
   - Reduce deforestation, and thereby stem the decline of photosynthesis caused by trees.
   - Reduce thawing of tundra, and thereby reduce the amount of methane that is released when tundra thaws.
   - Reduce thawing of polar ice, and thereby reduce the amount of methane that is released when polar ice melts.

Relevant Laws and Treaties

1. The **Kyoto Protocol (1997)** is an agreement among 150 countries to reduce greenhouse gases.

2. The **Clean Air Act (1972)** establishes carbon dioxide as a criteria pollutant and sets standards for it and other greenhouse gases.

Biodiversity

Biodiversity reflects the vastly different types of life on Earth. There are three types of biodiversity: genetic diversity, species diversity, and ecosystem diversity.

Genetic Diversity

Genetic diversity refers to the variety of different genes within the gene pool of a population of a single species. The importance of genetic diversity is based on an understanding of the role of genes and how they are expressed.

1. **Genes** are portions of DNA that contain the information needed to produce specific proteins. There is a separate gene for each protein, but there may be many different versions of a gene within a population of a single species. Some genes produce proteins that are more useful than others.

2. **Gene pool** is the collective group of traits that exist in all the chromosomes of all individuals in a population. It is like the cumulative library of genes—expressed and nonexpressed—that occur in a population. While one set of genes may prove useful at one time, another set of genes may be more useful under different conditions. Therefore, having a broad gene pool helps an overall population to survive under many different kinds of conditions.

3. **Founder effect** occurs when a small group of individuals "found" or start a new population. The overall gene pool of this new population is limited to what is available among the genes of the few founders. Therefore, as the population grows, its overall gene pool is restricted and reduces the survivability of that population.

4. **Bottleneck effect** is similar to the founder effect, but is created by the survival of a few individuals after some cataclysmic event. The survivors then "found" a new population with a low level of genetic diversity.

5. **Genetic assimilation** threatens some species when they crossbreed with closely related, more hardy species.

6. **Low genetic diversity** is most often seen when a few survivors are saved, or kept in zoos, to reestablish new populations. When a species hits a low enough level of genetic diversity so that extinction is an eventuality, it is said to have hit its lowest viable population size.

Species Diversity

Species diversity refers to the abundance of different types of species that are available.

1. **Endangered and extinct species**

    Endangered species are those at risk of extinction. An extinct species is a species that no longer exists.

2. **Traits of endangered species**

    - **Logistic population growth strategy** is described in Chapter 6. Individuals with a long life span who need a lot of resources per individual, have a large body weight, are high on the food chain, or have a low reproductive rate are the first species to be endangered. Logistic populations put more investment of resources into each individual. When the individuals perish, they do not replenish themselves as quickly and are more easily endangered.

    - **Large requirement for land** could mean that either the species is solitary and may not find mates, requires a lot of resources, or—because of migration—may require more than one ecosystem for its survival.

    - **Specialist species with a narrow niche** means that if the niche shifts—for example, through climate change or pollution—that organism would not be hardy enough to survive in the new niche.

    - **Low genetic diversity** was already explained in this chapter. Most particularly, species that are part of a genetic bottleneck, or experience genetic isolation or genetic assimilation are more easily threatened.

    - **Competition with a more dominant species,** such as humans, results in the more dominant species using the resources rather than the endangered species. One example is our use of old-growth forests for logging; we compete for the timber with forest species who depend on the timber for their habitat.

    - **Low tolerance for pollution** may cause an organism to be overpowered by a change in its surrounding abiotic conditions. This may be especially true for top carnivores who would be sensitive to the biomagnifications of toxins moving up the food chain.

    - **Strongly hunted species**—whether for some economically valuable commodity from the species, or because they are thought of as a nuisance—is another reason why some species are more at risk of extermination. Some examples include elephants (for their ivory), the North American grey wolf (to remove them as a predator of livestock), whales (for oil and other products), and sharks (for traditional medicines).

> **TEST TIP**
>
> Be sure to make the connection between the traits of endangered species, outlined above, and related population ecology topics found in Chapter 6. The knowledge will serve you well on the exam.

3. **Minimum viable populations**

    If the number of individuals in a breeding population drops below a critical value, then sexual reproduction of the adults will no longer provide the recombination of genes that is necessary to have the population survive over time. While the species may continue for several more generations, it is only a matter of time until the genetic diversity is not sufficient enough to provide the gene pool necessary to adjust to environmental changes in the surroundings.

4. **Benefits of species diversity**

    An ecosystem benefits from having a broad diversity of species. For example, if there are a number of different producers in a biological community, a change in climate and elimination of one producer may not pose as high a threat to all the other species in the community, than if the whole community depended on a single producer.

    Humans also benefit from the existence of broad species diversity. Much of our existence depends on adequately functioning ecological processes. The soil formation and nutrient cycling that we need to grow crops depend on microorganisms.

5. **Threats to species diversity**

    Species can become extinct because of natural reasons, as well as human causes. Natural extinctions may occur from climate changes that result from cataclysmic events, such as volcanic activity or asteroid impact, or from long-term changes, such as continental drift.

    Humans cause extinctions from a level of overhunting that outpaces the species' ability to reproduce. The most severe case of human-induced extinction is from habitat destruction or fragmentation, where humans denude or develop an area and remove the habitats of species—sometimes even without knowing that an endangered species has been affected.

    Sometimes human-induced extinctions occur because a new species has been introduced that outcompetes the indigenous species. For example, kudzu vines were introduced in the 1930s by the U.S. Soil and Conservation Service to control erosion. By now, this pernicious Japanese vine has taken over and eliminated natural flora in large portions of the Eastern United States.

Kudzu was purposefully introduced and carried an unintended outcome, but some species are inadvertently spread long distances in this age of global business and jet travel. One example is the Asian long-horn beetle, which was probably brought to the United States on wooden shipping pallets or products. The beetles burrow into trees, inhibit the flow of sap, and eventually kill trees.

Ecosystem Diversity

Ecosystems diversity refers to the abundance and distribution of different types of ecosystems around the globe. Ecosystems evolve and continue to shift over many millennia. They steadily provide essential habitats for a wide range of species upon which we depend. Diversity of ecosystems is important to support a variety of habitats necessary to support the species that depend on them.

Ecosystem biodiversity is connected to genetic and species diversity in that sufficient habitat is needed to sustain large enough populations so that there is significant genetic recombination within each generation. If the habitat is not large enough to sustain breeding, the number of genes passed to the next generation decreases, and the population is less able to adapt to changes in the environment. The Florida panther is an example of a drop in species diversity that has come from a drop in genetic diversity, which has in turn come from a decrease in ecosystem diversity. Additionally, a habitat must provide enough food for vulnerable populations. For example, decreased bamboo forests in China have endangered the survival of the panda. (See a review of terrestrial and aquatic biomes and ecosystems in Chapter 5.)

Protecting Biodiversity

Many of the management techniques mentioned in Chapter 7 also have a significant impact on biodiversity. Some of the specific biodiversity-related techniques are outlined below. Habitat loss is the largest reason for species extinction. Most of these techniques serve to protect habitats and, therefore, promote all three types of biodiversity.

1. **Reduce fragmentation.** Fragmentation of a population occurs when human development has turned a large contiguous ecosystem into a patchwork of subpopulations that are unable to interbreed, or have limited opportunity to roam.

When population breeding is restricted, genetic isolation or bottlenecks occur. Fragmentation also amplifies the *edge effect*—those effects experienced at the borders of any ecosystem. For example, the edge of a forest has greater light penetration than the core of the forest. As a result, different ground shrubbery and fauna prevail near the edge. Reducing fragmentation protects all three levels of biodiversity.

2. **Gap analysis** by policymakers helps to identify ways to cushion fragmented populations and establish public land to buffer core populations. For example, many national parks and wilderness areas are often surrounded by Forest Service land or Bureau of Land Management (BLM) land. These lesser-protected spaces buffer the more guarded park and wilderness lands.

3. **Corridors** between fragmented populations serve as genetic bridges between populations that might otherwise be isolated by roads and other human development. Corridors also provide an escape route for endangered populations in the event of a local disaster, such as fire or flood.

4. **Wetland protection** through protecting flood plains, minimizing channelization, minimizing lock and dam systems, decreasing beach erosion, preventing groundwater pollution, and refilling previously drained wetlands all protect this essential biome.

5. **Range management** helps prevent overgrazing of public lands and desertification of grasslands. Range leases for public lands can be issued at a rate that allows the land to replenish itself between grazing seasons.

6. **Debt-for-nature swap** allows financially destitute countries to trade development-related debt for an agreement to hold large tracts of land in reserve for wildlife and habitat preservation.

7. **Fire management** has vastly improved in the last decade or so. There are drawbacks to both a let-burn policy and a policy that immediately extinguishes fire in public lands. The current trend is to allow fires to burn except when they threaten life or property. Letting an area burn reduces ground-level fuel and maximizes an area's ability to recover biologically.

8. **Land reclamation** is required after land has been used for mining since the enactment of the Surface Mining Control and Reclamation Act (SMCRA). This law establishes state control of the process of reclamation, and requires mining operations to put funds in trust to pay for postmining reclamation back to a biologically useful habitat.

9. **Land use mitigation** is a type of trade for developers that allows them to purchase and protect one tract of land in exchange for developing another tract of land. For example, coastal condominiums might be built near the beach, but developers can establish and protect wetlands nearby in an area that would be ecologically equivalent, but less commercially desirable.

10. **Captive breeding programs and zoos** provide an essential rescue process for species on the brink of extinction. Although a species may never regain the genetic diversity it once enjoyed, breeding in captivity may allow further generations of an endangered species to be reintroduced into the wild to establish new populations.

Relevant Laws and Treaties

The **Endangered Species Act (1973)** protects the habitats of endangered species; limits harvesting, transporting, or selling endangered species; and limits commercial use of habitats occupied by endangered species.

See also the description of public lands that is provided in Chapter 7. These public lands—particularly national parks, national forests, wilderness areas, and wildlife refuges—are a major contributor to the conservation of ecosystem biodiversity.

Time for a quiz
- Review strategies in Chapter 2
- Take Quiz 7 at the REA Study Center
  (www.rea.com/studycenter)

Take Mini-Test 2
on Chapters 7–10
Go to the REA Study Center
(www.rea.com/studycenter)

Practice Exam

Also available at the REA Study Center *(www.rea.com/studycenter)*

> This practice exam is available at the REA Study Center. Although AP exams are administered in paper-and-pencil format, we recommend that you take the online version of the practice exam for the benefits of:
> - Instant scoring
> - Enforced time conditions
> - Detailed score report of your strengths and weaknesses

Practice Exam
Section I

(Answer sheets appear in the back of the book.)

TIME: 90 minutes
100 multiple-choice questions

Part A

Directions: Each set of lettered choices below refers to the numbered questions or statements immediately following it. Select the one lettered choice that best answers each question or best fits each statement and then fill in the corresponding oval on the answer sheet. A choice may be used once, more than once, or not at all in each set.

Questions 1–4 refer to the following population concepts. Select the one lettered choice that best fits each statement.

(A) r-strategist

(B) k-strategist

(C) ENSO

(D) demographic transition

(E) Coriolis effect

1. Tends to be less limited by environmental resistance

2. Tends to be more susceptible to extinction

3. Moves a population's age-structure diagram from pyramidal to more block-shaped

4. Decreases anchovy harvests

Questions 5–8 refer to the following types of gases. Select the one lettered choice that best fits each statement.

(A) nitrogen gas

(B) sulfur dioxide

(C) ozone

(D) radon

(E) volatile organic compounds

5. Major indoor pollutant with nonanthropogenic source

6. NOT a criteria pollutant

7. Major source is from coal-fired power plants

8. Irritant formed as a result of photochemical oxidation

Questions 9–12 refer to the following biogeochemical cycles. Select the one lettered choice that best fits each statement.

(A) carbon cycle

(B) nitrogen cycle

(C) phosphorus cycle

(D) water cycle

(E) sulfur cycle

9. Includes a component of proteins that eventually contributes to acid rain

10. The only cycle in which the element remains in the same oxide form

11. Sequestered in coral reefs and polar ice

12. Most closely associated with weather

Questions 13–16 refer to the following answers. Select the one lettered choice that best fits each statement.

(A) sequestration

(B) debt-for-nature swap

(C) fragmentation

(D) gap analysis

(E) corridors

13. Describes the storage of carbon in the form of forests or limestone   A

14. The practice of surrounding a national park with a national forest is an example   D

15. Turning a large contiguous population into clusters of smaller populations   C

16. A process that increases genetic biodiversity by allowing subpopulations to interbreed   E

Questions 17–20 refer to the following biomes. Select the one lettered choice that best fits each statement.

(A) taiga

(B) tundra

(C) temperate deciduous forests

(D) savannah

(E) cloud forests

17. Tend to occur on the windward side of equatorial mountains   E

18. The driest of these biomes   B

19. Northernmost forests   A

20. Lose their leaves in the fall   C

Questions 21–24 refer to the following zones. Select the one lettered choice that best fits each statement.

(A) euphotic zone

(B) limnetic zone

(C) pelagic zone

(D) benthic zone

(E) littoral zone

21. The section of a freshwater lake that is near the surface, where algae might grow A

22. The bottom of either a lake or an ocean D

23. The open water section of a lake where the sun does not easily penetrate B

24. The shallow portion of a lake, where rooted vegetation can grow E

Questions 25–28 refer to the following answers. Select the one lettered choice that best fits each statement.

(A) water diversion projects

(B) channelization

(C) fracking

(D) swidden

(E) purse seining

25. Used by indigenous people to clear small plots of land D

26. Blamed for salinization A

27. Cause of groundwater pollution C

28. Allows for navigation of larger boats and more flooding B

Part B

Directions: Each of the questions or incomplete statements below is followed by five suggested answers or completions. Select the one that is best in each case and fill in the corresponding oval on the answer sheet.

29. Coal-fired power plants generate which of the following types of pollution?

    I. Oxygen-depleting thermal water pollution
    II. Particulate air pollution
    III. Heavy metal air pollution

    (A) I only
    (B) II only
    (C) III only
    (D) I and II only
    (E) I, II, and III

30. In the figure shown below, the base of the pyramid harbors more energy within an ecosystem than the top of the pyramid. Which of the following factors explains why much of the energy from lower on the pyramid does NOT make it to the top?

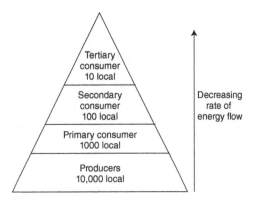

Ecological Energy Pyramid

(A) Energy is not conserved in ecosystems.
(B) Larger organisms at the top need less energy.
(C) Waste heat and unconsumed prey use up energy.
(D) Species extinction
(E) Species migration

31. Which of the following conditions in a river created the effect summarized by the chart below?

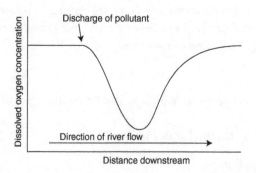

(A) Acid deposition from a nearby power plant has gone into the river.

(B) A sewage treatment plant disposed of nitrate-rich effluent into the river.

(C) A rainstorm caused water to wash over a parking lot and into the river.

(D) Oil from a tug boat washed into the river.

(E) Pesticides from a local feed-lot washed into the river.

32. Which location on the map below experiences decreased upwelling and increased sea surface temperatures during an El Niño event?

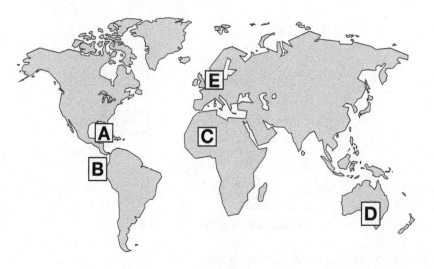

(A) A
(B) B
(C) C
(D) D
(E) E

33. Which of the following are considered "criteria pollutants"?

   I. Oxides of carbon and nitrogen
   II. VOCs
   III. Oxygen-demanding wastes in water

   (A) I only
   (B) II only
   (C) III only
   (D) I and II only
   (E) I, II, and III

34. Which of the following techniques measures toxic effects directly on living tissue?

   (A) Secchi disc
   (B) Chromotography
   (C) Tissue culture
   (D) Bioassay
   (E) Bioremediation

35.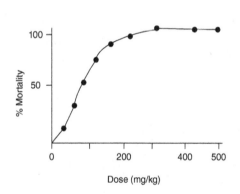

   Use the dose-response curve shown above to determine the LD_{50} of the toxin represented by the curve.

   (A) 50 mg/kg
   (B) 100 mg/kg
   (C) 200 mg/kg
   (D) 400 mg/kg
   (E) 500 mg/kg

36. Thermohaline currents in the ocean are caused by

   I. Temperature gradients
   II. Salinity gradients
   III. pH gradients

   (A) I only
   (B) II only
   (C) III only
   (D) I and II only
   (E) I, II, and III

37.

Dose (mrem)	Cause	Effect
3 mrem	5-hour plane flight	None
7 mrem/yr	Building materials	None
50 mrem	Diagnostic X-ray	None
1,000 mrem/yr	Safety threshold	None
10,000 mrem/yr		Cancer risk
30,000 mrem		Decreased white blood cells
350,000 mrem		Half die in 30 days

Considering the chart above, what is the ratio of the dose at which point an individual is at risk of cancer to the safety threshold?

   (A) 10,000:1
   (B) 1000:1
   (C) 100:1
   (D) 50:1
   (E) 10:1

38. In the aftermath of Hurricane Katrina, homes that were flooded sat in a high humidity environment. Many of these homes had to be destroyed because of the risk of

   (A) carcinogens
   (B) allergens
   (C) neurotoxins
   (D) teratogens
   (E) criteria pollutants

39. Which of the following represent mechanisms that form sedimentary rock?

    I. Mechanical weathering
    II. Chemical weathering
    III. Inclement weather

    (A) I only
    (B) II only
    (C) III only
    (D) I and II only ← circled
    (E) I, II, and III

40. Which of the following represents an example of a teratogen?

    (A) Criteria pollutants
    (B) Oxygen-demanding wastes
    (C) Alcohol during pregnancy; thalidomide ← circled
    (D) Mercury and other neurotoxins
    (E) Mine tailings

41. Which of the following Earth "spheres" contain molten magma?

    (A) Troposphere
    (B) Lithosphere ← circled
    (C) Biosphere
    (D) Hydrosphere
    (E) Stratosphere

42. Which of the following is a factor that contributes to the lethal nature of any toxin?

    I. Amount of dose
    II. Genetic predisposition
    III. Persistence

    (A) I only
    (B) II only
    (C) III only
    (D) I and II only
    (E) I, II, and III ← circled

43. Which of the following is the location of most of the Earth's weather?

    (A) Troposphere
    (B) Stratosphere
    (C) Antarctic
    (D) Jet stream
    (E) El Niño

44.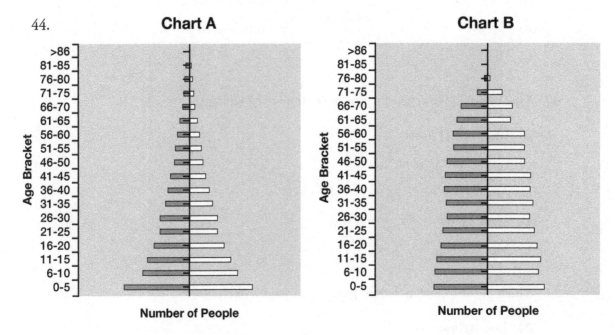

The above histograms show the age-structure diagram of the same population at different moments in its history. Which of the following is a likely catalyst for the change documented in this population data?

    (A) Population decreased through more birth control.
    (B) Migration of older people out of the country has ceased.
    (C) The land has been degraded, creating greater poverty.
    (D) Wealth increased, allowing for greater longevity.
    (E) Infant mortality increased.

45. Which of the following forms of energy is NOT ultimately caused by the energy from the sun?

    (A) Hydroelectric
    (B) Fuel oil
    (C) Natural gas
    (D) Geothermal
    (E) Passive solar

46. Solar energy may be captured in various ways, ranging from using photovoltaic cells to simply storing evaporated water. The type of solar heating that occurs as a result of the design of a building, such as a trombe wall, extended eves, or planting deciduous trees on the sunny side, is collectively known as

   (A) passive solar heating
   (B) active solar heating
   (C) photovoltaic energy
   (D) greenhouse effect
   (E) leapfrog effect

47.

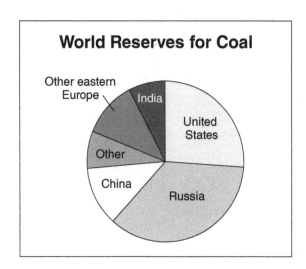

Consider the pie chart above. What approximate percent of the world reserves is represented by the United States and Russia?

   (A) About 20%
   (B) About 40%
   (C) About 60%
   (D) About 80%
   (E) Nearly 100%

48. Which of these represents the most sustainable, least energy-intensive method of making sure crops receive enough water?

   (A) Using water from the aquifer for irrigation.
   (B) Creating a water diversion project to transport water.
   (C) Using reverse osmosis to convert seawater to fresh water.
   (D) Planting crops that survive on the rain where they are planted.
   (E) Using a drip system to irrigate crops with water pumped from a river.

49. Which answer most closely represents the relationship between climate change and ocean acidification?

    (A) Warmer ocean waters allow more hydronium ions to dissolve.

    (B) Atmospheric carbon dioxide increases ocean levels of carbonic acid.

    (C) Warmer air above the sea draws acid ions out of marine organisms.

    (D) Atmospheric carbon dioxide causes ocean pH to increase.

    (E) Warmer ocean water promotes higher metabolism among marine organisms.

50. Which of the following accelerates the impact of the greenhouse effect on climate change?

    I. Deforestation
    II. Melting of arctic ice
    III. Rising sea levels

    (A) I only
    (B) II only
    (C) III only
    (D) I and II only
    (E) I, II, and III

51.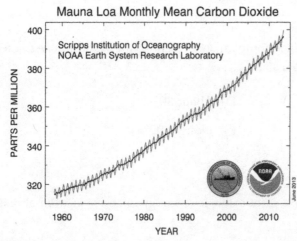

    The data summarized above was obtained from NOAA from an observatory in Hawaii. How does the effect measured by this data potentially influence climate change?

    (A) An increase in a greenhouse gas over time may help retain heat near the Earth.

    (B) There is no connection between the data on this chart and climate change.

    (C) Decrease in polar reflectivity will accelerate global warming.

    (D) The annual cycles of carbon dioxide are increasing in magnitude.

    (E) Atmospheric carbon dioxide has decreased over time.

52. Ozone depletion is most likely caused by

    (A) mercury
    (B) oxides of nitrogen
    (C) chlorofluorocarbons
    (D) carbon dioxide
    (E) methane

53. Which type of remediation uses quickly adapting living organisms to turn toxic wastes into non-toxic products?

    (A) Bacterial remediation
    (B) Chemical remediation
    (C) Physical remediation
    (D) Sterilizing remediation
    (E) Artificial remediation

54. After all other forms of remediation have been tried, which of the following methods is often used to isolate toxic waste from further exposure by burying it?

    (A) Reclamation
    (B) Mine tailing
    (C) Capping
    (D) Encapsulation
    (E) Filtration

55. Which of the following processes vaporizes underground toxic spills, such as those created from ruptured oil pipelines?

    (A) Bioremediation
    (B) Air stripping
    (C) Distillation
    (D) Vapor retribution
    (E) Soil flushing

56. Which of the following is an example of a point-source pollutant?

    (A) Forest fires
    (B) Agricultural application of pesticides
    (C) Oil waste washed off parking lots during a rainstorm
    (D) Sewage treatment plants
    (E) Air pollution from automobiles

57. A lake process marked by high levels of nutrients and eventually turns the lake first into a bog and then solid land is known as

   (A) oligotrophication
   (B) eutrophication
   (C) stratification
   (D) photosynthesis
   (E) spring turn-over

58. Which of the following is most likely to result when the banks of a river are cemented into a fixed channel?

   (A) Increased river velocity, erosion, and flooding
   (B) Increased pollution
   (C) Decreased macrobiotic life in the river bed
   (D) Decreased runoff from rain
   (E) Decreased bank erosion and flooding

59. The ecological principle in which one biological community replaces another is known as

   (A) evolution
   (B) edge effect
   (C) mutualism
   (D) succession
   (E) bottleneck effect

60. Which of the following is caused by an El Niño Southern Oscillation (ENSO) event?

   (A) Warmer water off the west coast of South America
   (B) Warmer water off the east coast of South America
   (C) Cooler water off the west coast of South America
   (D) Increased global warming
   (E) Increased melting of Greenland ice

61. Settling of a ground surface due to overdrawing water from nearby wells is called

   (A) upwelling
   (B) subsidence
   (C) vertical stratification
   (D) plume
   (E) downburst

62. Stone fly nymphs live in beds of streams that contain a high level of dissolved oxygen. When the dissolved oxygen decreases in the stream, stone fly nymphs can no longer live there. Stone fly nymphs are an example of

    (A) a keystone species
    (B) an indicator species
    (C) an endangered species
    (D) a marine species
    (E) a threatened species

63. Which of the following represents the smallest use of freshwater?

    (A) Residential
    (B) Agricultural
    (C) Paper and pulp mills
    (D) Industrial uses
    (E) Cooling power plants

64. Which of the following occurs as a result of acid rain?

    (A) Global warming
    (B) Ozone depletion
    (C) Aluminum poisoning in plants
    (D) Increase in stream pH
    (E) Increase in cancer

65. Which of the following is an adaptation that plants use in hot, dry regions?

    (A) Broad leaves
    (B) Waxy leaves and thorny stems
    (C) Thin leaves on plants that are low to the ground
    (D) Broad, shallow web of roots
    (E) High leaf-to-woody-stem ratio

66. Which technique of air pollution remediation turns an air pollution problem into a solid waste problem?

    I. Fluidized bed combustion
    II. Electrostatic precipitator
    III. Catalytic converter

    (A) I only
    (B) II only
    (C) III only
    (D) I and II only
    (E) I, II, and III

67. Which of the following is an initial step in accomplishing a demographic transition?

    (A) Decrease in death rate
    (B) Decrease in birth rate
    (C) Increase in birth rate
    (D) Increase in death rate
    (E) No change in either birth or death rate

68. The epicenter of an earthquake exists

    (A) on the surface of the Earth
    (B) at the actual spot underground where two tectonic plates adjust position relative to one another
    (C) under the ocean
    (D) where the earthquake does the most damage
    (E) at the position on the Earth directly opposite where the quake is felt

69. A seismically active arc of islands, such as Alaska's Aleutian Island Chain, is an indicator of

    (A) the convergence of oceanic and continental plates
    (B) a mid-plate hotspot
    (C) the convergence of two oceanic plates
    (D) a divergent boundary
    (E) a mid-ocean rift

70. How long does it typically take to produce one millimeter of topsoil?

    (A) One week
    (B) One year
    (C) One decade
    (D) One hundred years
    (E) One month

71. Most of the initial oxygen in the early atmosphere was created by

    (A) volcanoes
    (B) cyanobacteria
    (C) red algae
    (D) plants
    (E) animals

72. Which of the following is measured in the number of deaths per 1,000 individuals in a population?

   (A) Fecundity
   (B) Fertility
   (C) Morbidity
   (D) Mortality
   (E) Natality

73. A lizard population of 5,000 experienced a crude birth rate of 60 and a crude death rate of 10. What is the total annual growth rate of this population?

   (A) Under 1%
   (B) Between 1% and 3.5%
   (C) Between 3.5% and 6%
   (D) Between 6.1% and 9%
   (E) Over 9%

74. What is the doubling time of a population that grows at a 3.5% annual growth rate?

   (A) 1 year
   (B) 3.5 years
   (C) 5 years
   (D) 10 years
   (E) 20 years

75. The following legislative action or international agreement works to limit the use and production of greenhouse gases.

   (A) SMCRA
   (B) NEPA
   (C) CERCLA
   (D) Montreal Protocol
   (E) Kyoto Protocol

76. This is the only type of federal public land that attempts to achieve the greatest good for the greatest number of people. Sometimes, these lands carry the sign, "A land of many uses."

   (A) Wilderness areas
   (B) Wildlife refuges
   (C) Bureau of Land Management
   (D) National Forest Service
   (E) National Park Service

77. Which of these river zones represents the intersection between aquatic and terrestrial biomes?

(A) Decomposition zone
(B) Septic zone
(C) Recovery zone
(D) Riparian zone
(E) Channel zone

78. A transfer of energy involves three steps, each of which is 80% efficient. What is the approximate total efficiency of all three steps?

(A) 24%
(B) 51%
(C) 66%
(D) 80%
(E) 98%

79. Which of the following supplies the greatest amount of energy in developed countries?

(A) Oil
(B) Coal
(C) Natural gas
(D) Hydroelectric
(E) Solar

80. A population experienced a birth rate of 23 people per thousand, and a death rate of 11 people per thousand. What is the approximate growth rate of the population?

(A) 0.12%
(B) 1.2%
(C) 2.4%
(D) 3.8%
(E) 4.2%

81. A stream with pH of 4.0 is

(A) strongly acidic
(B) weakly acidic
(C) neutral
(D) weakly basic
(E) strongly basic

82. In the rock cycle, metamorphic rocks are formed

    (A) during the weathering process, as sediments pile on top of each other
    (B) when magma solidifies
    (C) from both sedimentary and igneous rocks through intense heat and pressure
    (D) only from sedimentary rock, when it melts and reforms
    (E) only from igneous rock, when it erodes

83. Denitrifying bacteria converts

    (A) nitrates into atmospheric nitrogen
    (B) atmospheric nitrogen into ammonia
    (C) ammonia into nitrites
    (D) nitrites into nitrates
    (E) nitrites into sulfates

84. The underground section between the surface of the Earth and a water table is called the

    (A) recharge zone
    (B) confined aquifer
    (C) transform boundary
    (D) zone of aeration
    (E) divergent boundary

85. The half-life of a radioactive isotope is 8 days. What percentage of the sample remains after 32 days?

    (A) approximately 1.5%
    (B) approximately 6%
    (C) approximately 12%
    (D) approximately 25%
    (E) approximately 50%

86. A population takes about seven years to double in size. This population's annual percentage growth is approximately

    (A) 2%
    (B) 7%
    (C) 10%
    (D) 15%
    (E) 22%

87. Which of the following is NOT a trait demonstrated by k-strategist organisms?

    (A) Population tends to overshoot the carrying capacity
    (B) Adults invest considerable energy in the next generation
    (C) Higher on the food chain
    (D) Tend to be long-lived
    (E) Tend to mature slowly

88. Which of the following is the most prevalent form of water pollution?

    (A) Oxygen-demanding wastes
    (B) Pathogens
    (C) Heavy metals
    (D) Sediment
    (E) Acid deposition

89. A 200 W bulb shines for 4 hours. How much energy is used?

    (A) 0.8 kWh
    (B) 8.0 kWh
    (C) 80 kWh
    (D) 40 kWh
    (E) 20 kWh

90. 100 grams of water are heated from 60°C to 80°C. How much energy is used to accomplish this? (C_{water} = 4.2 J/g°C)

    (A) 84 J
    (B) 840 J
    (C) 8,400 J
    (D) 84,000 J
    (E) 8.4 J

91. If a 100 W light bulb is 80% efficient. What amount of heat energy is given off over a 100 hour period?

    (A) 20 W
    (B) 80 W
    (C) 2 kWh
    (D) 20 kWh
    (E) 8 kWh

92. Sulfur dioxide has an EPA standard set at 75 ppb. What would the standard be in terms of *parts-per-million*?

   (A) 750.0 ppm
   (B) 75.0 ppm
   (C) 7.5 ppm
   (D) 0.75 ppm
   (E) 0.075 ppm

93. Integrated pest management (IPM) involves the use of

   I. Pesticides
   II. Natural predators
   III. Sex attractants to lure insects into traps

   (A) I only
   (B) II only
   (C) III only
   (D) I and II only
   (E) I, II, and III

94. The environmental consequences of raising beef in feedlots include

   I. Nutrient pollution of groundwater and nearby waterways
   II. Increased bacterial resistance to antibiotics
   III. Increased levels of hormones in our food

   (A) I only
   (B) II only
   (C) III only
   (D) I and II only
   (E) I, II, and III

95. Which of the following is NOT a renewable source of energy?

   (A) Organic material compressed to form coal
   (B) Corn crops fermented to form ethanol
   (C) Solar energy
   (D) Hydroelectric power
   (E) Tidal energy

96. The reason that the Earth experiences *seasons* is most closely related to

    (A) the distance between the sun and the Earth at certain times of the year
    (B) the position of the moon between the Earth and the sun, thus blocking sunlight
    (C) the tilt of the Earth's axis
    (D) whether a portion of the Earth faces toward or away from the sun
    (E) the amount of global warming that has taken place

97. Normally, rain is slightly acidic due to

    (A) sulfurous acid created from coal combustion
    (B) metallic acid deposition from mining wastes
    (C) carbonic acid created by carbon dioxide in the atmosphere
    (D) nitrous acid created from oxides of nitrogen mixed with water
    (E) a combination of anthropogenic sources

98. Which of the following is typically used to remove or neutralize waterborne pathogens in the water treatment process?

    I. UV light
    II. Ozone
    III. Soap

    (A) I only
    (B) II only
    (C) III only
    (D) I and II only
    (E) I, II, and III

99. Ozone in the stratosphere is important to life on Earth because it

   (A) blocks infrared radiation

   (B) absorbs microwaves

   (C) breaks down toxic chlorofluorocarbons

   (D) blocks ultraviolet light

   (E) absorbs greenhouse gases

100. Which of the following does NOT represent pronatalist pressure on a family who lives in a country that has not yet undergone a demographic transition?

   (A) Status of being wealthy and fertile enough to have children

   (B) Educational and occupational opportunities for parents

   (C) Have enough children so that some survive into adulthood

   (D) Children are a source of cheap labor.

   (E) Children are needed to support parents in old age.

Section II
Free-Response

TIME: 90 minutes

4 free-response questions

Directions: You have 90 minutes to answer all four of the following questions. Because each question will be weighted equally, it is suggested that you divide your time equally among the questions. Each answer should be organized, well balanced, and as comprehensive as time permits.

Answers must be in organized well-written prose form; outline form is *not* acceptable. Do not restate the questions. If a specific number of examples are called for, no credit will be given for additional examples.

Write you answers with a pen only. Write clearly and legibly. If you make an error, cross it out rather than trying to erase it.

1. Answer the questions that relate to the following article.

Developers Eager to Create Habitat

In recent weeks, Johnson and Smith, LLC, a major Bay Area developer of residential neighborhoods, has been jumping at any chance to buy land in the Fremont city limits. That is nothing new, but this is different because they are not going to use the land to build houses, but to preserve a pristine habitat for endangered frogs, foxes, and fish.

The beneficiaries are the California red-legged frog, the Delta smelt, and the San Joaquin fox. The U.S. Fish and Wildlife Service has stated that developers will be allowed to use the habitat of these species for housing projects if they offset the damage by preserving a place for these species somewhere else.

Critics of this arrangement state that this will create an active real estate market for land that needs to be left alone. A representative from Johnson and Smith responded by saying that the land will be put into a trust once it is purchased. This will keep the land off the secondary real estate market.

(continued)

> This type of exchange has prompted other developers to begin buying ecologically significant land that previously had no commercial value, such as wetlands. Tad Pole, owner of 320 acres of wetland near the delta, states, "This is a real windfall for me. I could not have afforded to preserve these lands. Now someone is coming to me with an offer that is many times what I could have gotten on the market in previous years. Now I'm thinking of retiring."
>
> While Mr. Pole and several species of wildlife are probably feeling pretty good about this, some developers complain that this arrangement increases costs that they have to pass on to the buyers of the homes they build. "Well, at least everyone is getting a place to live," adds Mr. Pole.

(a) Identify a major law that is being enforced by the U.S. Fish and Wildlife Service. What is the primary effect that this law has on biodiversity?

(b) What is the commonly used term for the arrangement described in this article?

(c) Use your understanding of environmentally healthy full-cost analysis to respond to the developer's complaint about having to pass on the cost of the preserves to home buyers.

(d) Of the species mentioned, pick one and identify three traits of the species that predispose it to be rare or endangered.

(e) Even though land has been set aside for wildlife, it doesn't necessarily mean that it is safe from environmental hazards. Pick three remaining risks to survival of these species. Briefly explain why you chose these risks.

2. The following information refers to Fremont's major coal-fire electricity generation plant.

Daily electrical output	13,000 MWh
BTU equivalent to 1.0 MWh	3,400,000 BTU
BTUs produce by 1.0 lb coal	5,000 BTU
Power plant overall efficiency	36%

(a) How many BTUs are produced by the plant each day?

(b) How many pounds of coal are used by the plant each year?

(c) What amount of energy is lost as heat each day, in BTUs?

(d) Assume that 80% of the waste heat is absorbed by the nearby river, which is diverted to cool the power plant. Local regulations limit the utility to a 5°F increase in the temperature of the river.

    i. Describe the environmental consequences of using the river water as a coolant for the power plant.

    ii. What weight of water, in pounds, is needed to absorb one day's waste heat from the plant?

3. Use the data table to answer the following questions.

Measurement	Location A	Location B
DO (ppm)	10.0	3.0
Temperature (C)	18.0	27.4
pH	7.5	6.0
Nitrates (ppm)	2.0	6.0
Diversity (#)	16.0	7.0
Abundance (#)	525.0	24.0

(a) Which point on the stream, A or B, seems to demonstrate the greatest ecological health? Use the data to provide reasons for your answer.

(b) Which of the above measurements represent biotic factors in the stream environment?

(c) Which device would be used to measure dissolved solids? What device would be used to measure undissolved solids?

(d) Write a hypothesis about what may have caused the change in stream health between point A and point B. Support your hypothesis with data from the chart.

4. You have inherited a parcel of land in an undeveloped portion of a moderately sunny climate. You would like to build a home that is as self-sufficient as possible, eliminating as many toixc compounds as you can. The land is somewhat isolated—far from power, water, and sewer systems.

(a) You can bring electrical power to the home, but the utility says that you have to pay for the cable to bring the power from the nearest transmission line. For the price that you would have to pay the electric company, you could install your own alternative energy system. Identify four strategies that you could use to meet needs that are typically met by electrical energy from a utility.

(b) Every household needs to be able to eliminate septic waste from sinks, toilets, showers, and washing machines. Because you are too far from the town's sewer line, you have to install a septic system. Describe how the nitrogen cycle is involved with septic treatment systems.

(c) You are moving to the country to escape air pollution, but indoor air pollution in a rural home can be just as unhealthy. Identify three common indoor air pollutants and a method for reducing each pollutant in your home.

(d) You know that someday you may want to sell your home, so you'd like documentation of your efforts by having your home LEED-certified. Other than electricity, wastewater, and indoor air pollution, what is at least one other aspect of building your home that would reduce your environmental impact and help with your LEED certification?

(e) Identify 10 steps that you could take in the building and servicing of this home to reduce the home's carbon footprint in comparison with a home built using standard methods.

Answer Key

Section I

1. (A)	26. (A)	51. (A)	76. (D)
2. (B)	27. (C)	52. (C)	77. (D)
3. (D)	28. (B)	53. (A)	78. (B)
4. (C)	29. (E)	54. (C)	79. (A)
5. (D)	30. (C)	55. (B)	80. (B)
6. (D)	31. (B)	56. (D)	81. (A)
7. (B)	32. (B)	57. (B)	82. (C)
8. (C)	33. (D)	58. (A)	83. (A)
9. (E)	34. (D)	59. (D)	84. (D)
10. (C)	35. (B)	60. (A)	85. (B)
11. (A)	36. (D)	61. (B)	86. (C)
12. (D)	37. (E)	62. (B)	87. (A)
13. (A)	38. (B)	63. (A)	88. (D)
14. (D)	39. (D)	64. (C)	89. (A)
15. (C)	40. (C)	65. (B)	90. (C)
16. (E)	41. (B)	66. (D)	91. (C)
17. (E)	42. (E)	67. (A)	92. (E)
18. (B)	43. (A)	68. (A)	93. (E)
19. (A)	44. (D)	69. (C)	94. (E)
20. (C)	45. (D)	70. (B)	95. (A)
21. (A)	46. (A)	71. (B)	96. (C)
22. (D)	47. (C)	72. (D)	97. (C)
23. (B)	48. (D)	73. (C)	98. (D)
24. (E)	49. (B)	74. (E)	99. (D)
25. (D)	50. (D)	75. (E)	100. (B)

Detailed Explanations of Answers

Section I

1. **(A)**

   An r-strategist demonstrates irruptive growth that can easily overshoot the carrying capacity of an environment. A k-strategist is more sensitive to environmental resistance and, therefore, does not tend to overshoot an environment's carrying capacity. A k-strategist invests too much in each individual to waste them during a die-off as a result of overshooting the carrying capacity.

2. **(B)**

   A k-strategist is more susceptible to extinction because more is invested in each individual, so each individual carries a higher responsibility for survival. They tend to mature later, and so cannot quickly replenish the species if several have died off. They are more sensitive to environmental resistance, so changes in the environment can have a more devastating effect.

3. **(D)**

   A demographic transition creates a more block-shaped age-structure histogram, instead of a pyramidal shape, because birth rates decrease and the longevity of the older age brackets increases.

4. **(C)**

   ENSO, or El Niño Southern Oscillation, causes nutrient upwelling off the coast of Peru (essential for anchovy survival) to cease.

5. **(D)**

   Radon is a major indoor pollutant from a nonanthropogenic source. Radon is responsible for about half the lung cancer cases in the United States, second to smoking. Radon is produced in the Earth and in earthen materials, like concrete, during the decay of radioactive heavy elements. Radon is the only decay product that is a gas, so it bubbles out of the ground when it is formed, then gets lodged in a well-insulated home and resides there unless ventilation is adequate to remove it.

6. **(D)**

    Radon is the only gas on the list that is not a criteria pollutant. The criteria pollutants were designated by the Clean Air Act as the most important gases to regulate. They include the oxides of carbon, nitrogen, and sulfur; photochemical oxidants, such as ozone; heavy metals and halogens, such as lead, mercury, and CFCs; particulate matter; and volatile organic compounds.

7. **(B)**

    Sulfur dioxide mostly comes from coal-fired power plants. The reason why coal contains sulfur is related to how coal is formed. Most coal was created from anaerobic decomposition of plants. The disulfide linkages in the proteins within those plants break apart during decomposition, leaving sulfur in the coal. When the coal is burned, the sulfur in the coal is oxidized and combines with oxygen to form sulfur dioxide. When the sulfur dioxide reacts with water vapor in the atmosphere, it forms acid rain. Sulfur dioxide from coal combustion is responsible for about 70% of acid rain.

8. **(C)**

    Ozone is the irritant formed from photochemical smog. While ozone is necessary in the stratosphere to protect us from damaging UV light, it is a health hazard when we breath it in at ground level.

9. **(E)**

    The disulfide linkages that give proteins their three-dimensional shape are composed of sulfur. This sulfur remains when a plant dies, becomes compressed, and begins the process that eventually forms coal. When the coal is burned, the sulfur in the coal is oxidized to sulfur dioxide, which when combined with water vapor in the air, creates acid rain.

10. **(C)**

    The element phosphorous always remains in the phosphate form (PO_4^{-3}) during the entire phosphate cycle, even as it undergoes chemical changes with the other elements combined with it. In the nitrogen, carbon, and sulfur cycles, the principal element changes in the oxidation state and recombines with other elements through their cycles. It is true that water does not undergo a chemical change in its cycle, but there is no chemical change whatsoever during the water cycle, only changes in location and physical phases.

11. **(A)**

    Carbon is sequestered in coral reefs as the carbonate ion (CO_3^{-2}) and in polar ice as methane (CH_4).

12. **(D)**

    The water cycle is responsible for the heating and cooling associated with weather. For example, when a cool air mass hits a cloud of water vapor, water can condense and fall as rain.

13. (A)

Sequestration is a word that refers to storage. In this case, it refers to the storage of carbon in the form of limestone or forests. Forests store carbon during the process of photosynthesis, which converts carbon in the form of carbon dioxide into carbohydrates that make up a tree.

14. (D)

Gap analysis by policymakers helps to identify ways to cushion fragmented populations and establish public land to buffer core populations. Surrounding a national park with a national forest creates a buffer system for the species that might be protected in the park. This type of action comes about after considering the gaps in the policy of protecting a species with a park.

15. (C)

Genetic fragmentation occurs when a large population is divided into a number of smaller groups that can no longer interbreed. This process can easily occur when building roads, for example; it can create a genetic bottleneck within any of the resulting subgroups of the population.

16. (E)

Providing a corridor between one population and another is a solution to fragmentation. Such a corridor can reverse genetic isolation by allowing subpopulations to interbreed.

17. (E)

Cloud forests usually occur on the windward side of equatorial mountains. They receive almost constant rainfall or mist because the moist air releases water to decrease pressure enough to pass over the mountain.

18. (B)

Of the biomes listed, tundra is the driest. While it may have some water, most of it is bound into ice or permafrost. The next driest biome listed is a savannah, but savannahs tend to have a rainy season at some time during the year.

19. (A)

Taiga forests are the northernmost forests. Taiga forests border on both arctic and alpine tundras.

20. (C)

Deciduous trees lose their leaves in the fall.

21. (A)

The euphotic zone is the section of the lake near the surface where algae typically grow.

22. (D)

The bottom of the lake or ocean is called the benthos, and benthic organisms live there.

23. (B)

Underneath the euphotic zone in the middle of a lake is the limnetic zone, where light does not easily penetrate.

24. (E)

The littoral zone is the shallow portion of the lake, where rooted vegetation grows. This is an important nursery zone where small fish and invertebrate larvae grow to adulthood. It may also be the area where a moose walks through to eat the tender roots of littoral grasses.

25. (D)

Swidden agriculture has been used by indigenous peoples for centuries, where a patch of land is cleared in order to grow a variety of food crops, usually more than one kind of crop in any location. This method worked well enough at low population levels. However, as populations increase, there is not enough time for the land to become naturally restored, and land degradation has increased.

26. (A)

Large water diversion projects, such as with the Aral Sea in current-day Uzbekistan, often result in salinization of the source as well as the receiving location as it accepts increasingly saline water. As water is removed, evaporation exceeds water inflow, which concentrates dissolved ions in the water. When that water is diverted to be used on crops, the soil becomes increasingly saline and eventually unusable.

27. (C)

Hydraulic fracturing, or fracking, injects large amounts of chemicals into the groundwater to fracture hydrocarbon-containing underground fissures, helping to release the oil or gas so that it can be recovered with wells. Those chemicals remain underground and have the potential of seeping into groundwater.

28. (B)

Channelization is the process of dredging a river and making the bank more defined. This creates a channel that makes the river more useful for navigation. However, channelization removes wetlands that allow water to spill over and be impounded during higher levels of runoff, increasing the flow—and therefore flooding—when the river system must accommodate more water.

29. (E)

Unfortunately, all these types of pollution are generated by coal-fired power plants. Thermal water pollution is caused by warm water effluent from the cooling process in the plant. Fly ash causes particulate pollution, but is lessened by the use of scrubbers. The burning of coal releases mercury contained within the coal. Not mentioned in these answer options is the release of oxides of nitrogen and sulfur, both of which cause acid rain.

30. (C)

The Second Law of Thermodynamics states that some energy will be lost to waste heat with each energy transition. This waste heat, or entropy, along with uneaten or undigested food from prior trophic levels, creates the pyramid shape and ensures that less energy exists at the top of the pyramid than at the bottom, within any ecosystem.

31. (B)

Nitrates are a nutrient that an alga uses to grow. When the algae blooms, so do the decomposers that break it down. These decomposers use up the oxygen in the water. Therefore, the net effect of adding nitrates to water is to decrease the amount of oxygen in the water.

32. (B)

An El Niño event causes equatorial currents to reverse, moving warm water that has remained at the equator across the Pacific Ocean towards South America, rather than away from it. An El Niño event also causes upwelling on South America's western coast.

33. (D)

Criteria pollutants are air pollutants that are identified by the Clean Air Act. Oxygen-demanding pollution of water is not considered a "criteria pollutant."

34. (D)

A bioassay measures toxic effects directly on living tissues or nonhuman organisms.

35. (B)

LD_{50} is the median lethal dose for 50% of the population, as derived from a dose-response curve. In the case represented by this data, the LD_{50} is 100 mg/kg.

36. (D)

Thermohaline currents are created by temperature and salinity gradients, both of which cause water to have a different density. As a result, water in the ocean flows from high-density locations to lower-density locations, thus creating a current.

37. **(E)**

It requires a dose that is ten times the safety threshold in order to cause cancer. Therefore, the ratio is 10:1.

38. **(B)**

Mold, mildew, and dust are common allergens, which catalyze the body's allergic response.

39. **(D)**

Mechanical and chemical weathering are mechanisms that break down large rocks into smaller rocks, which then settle and become compressed into sedimentary rock.

40. **(C)**

Teratogens cause birth defects. Alcohol during pregnancy can cause birth defects, just as thalidomide is able to do, if taken during pregnancy.

41. **(B)**

The lithosphere is composed primarily of molten magma. The lithosphere exists just below the crust.

42. **(E)**

All of these factors—size of dose, genetic predisposition, and persistence (in the environment)—have the ability to increase the lethality of any toxic compound. Additionally, solubility may also play an important role in toxicity. For example, fat-soluble toxins increase the level of both biomagnifications and bioaccumulation.

43. **(A)**

Most of the Earth's weather, and most of the molecules in the atmosphere, occur in the troposphere, which is the layer of atmosphere closest to the Earth.

44. **(D)**

Increased wealth is a major catalyst for a demographic transition, followed by increased health care costs, increased social security, decreased infant mortality, and increased longevity.

45. **(D)**

One doesn't have to go very deep into the ground to find a constant temperature. This geothermal energy is ultimately fueled by the nuclear reactions in the center of the Earth, and, using a geothermal heat pump, can be tapped from any place on Earth. A geothermal heat pump is like a refrigerator that pumps heat from one place to another. When it is cold, a geothermal heat pump moves heat from the Earth to the home; when it is hot, it works in reverse. The other forms of energy listed are in some way fueled by the power of the sun, which powers photosynthesis and the evaporation of water, and so on.

Detailed Explanations of Answers 247

46. (A)

Passive solar heating refers to the type of solar heat that does not use energy and a pump at some point. It depends on non-movable design or placement of eves, windows, trees, storage masses, and orientation to maximize the amount of solar energy taken in.

47. (C)

The combination of the contributions of the United States and Russia compose just under two-thirds of the total amount, or about 60%.

48. (D)

Transporting water requires energy, which must be supplied and adds to the cost of producing a crop; the more energy used, the less sustainable the technique. Also, there are environmental consequences of taking water from one ecosystem and transporting it to another. The least costly, from an economic and environmental perspective, is watering crops from rain, without transport or pumps.

49. (B)

Through Henry's Law, the amount of carbon dioxide in the atmosphere is proportional to the amount of dissolved carbon dioxide in the sea. This amount of dissolved carbon dioxide in some small fraction combines with water to increase the amount of aqueous carbonic acid and to decrease the pH in the sea.

50. (D)

As the tundra thaws, it releases methane, which further contributes to global warming. Also, as forests are cut down or die, they sequester less carbon dioxide, which then remains in the atmosphere as a greenhouse gas. Both of these represent positive feedback loops that accelerate climate change. Rising sea levels are a result of climate change, but do not themselves accelerate further climate change.

51. (A)

The chart shows an increase in atmospheric carbon dioxide over the last several decades. Since carbon dioxide is a greenhouse gas, it suggests that more heat is being retained within the troposphere, thereby affecting climate.

52. (C)

Chlorofluorocarbons are catalysts that accelerate the conversion of atmospheric ozone to oxygen gas.

53. (A)

Bacterial remediation is extremely effective because bacteria adapt so quickly, new strains evolve to digest a toxin and convert it into nontoxic waste.

54. (C)

Capping simply places a cap on buried waste. Capping takes place in many forms. Sometimes a cap is simply a layer of clay; sometimes it is many tons of concrete.

55. (B)

Air stripping allows hydrocarbons that are already somewhat volatile to be more mobile. Air stripping contaminated soil removes volatile organic compounds, which either leave the soil and are emitted into the atmosphere or are collected and disposed of in some other way.

56. (D)

Point-source pollution comes from a single stack or pipe, and is remediated by changing the behavior at a single point. Sewage treatment plants put their effluent into a river or creek from a single location.

57. (B)

Eutrophication is a change in a lake that begins with an excess of nutrients. Algal blooms begin, die, and form a mat that littoral grasses use as a source of nutrients. The littoral zone becomes steadily more littered with organic material, and it evolves into a boggy area, and then a more solid shoreline. This process gradually decreases the size of the lake until it is filled in.

58. (A)

Channelization limits the expansion of a river during flood season. As a result, the velocity of the water increases, which exerts a greater force on the river bank and increases flooding.

59. (D)

Succession is the process by which one biological community yields to another, until a climax community is established. Primary succession occurs when land begins from bare rock. Secondary succession occurs when the land is stripped or damaged, and it returns to a formerly held climax community.

60. (A)

ENSO causes equatorial currents to reverse, causing warm water to pile up on the west coast of South America.

61. (B)

Subsidence is the settling of the ground when the subsurface ground has been compressed or chemically eroded.

62. (B)

A stone fly nymph is an indicator species because it provides an indication of oxygen levels in the stream. The nymphs, like other indicator species, provide qualitative evidence of an important environmental parameter without a human, for example, having to take a quantitative measurement.

63. (A)

Residential consumers represent the smallest number of users of freshwater of all the users mentioned. Agriculture is the largest user, and paper and pulp mills are the largest industrial users.

64. (C)

When acid rain falls on rocks and the ground, aluminum ions are leached out of the Earth, which are themselves acidic when in water, and damage plants and people.

65. (B)

Waxy leaves and thorny stems are two adaptations that help plants retain water. The waxy leaves help hold water in, and the thorny stems are a competitive scheme to repel animals who attempt to steal the water from the plant. Broad leaves are an adaptation under rainforest canopies among plants that compete for sun. Thin leaves low to the ground are tundra adaptations. Broad shallow roots are another rainforest adaptation, where the soil is thin and the few nutrients available are close to the surface.

66. (D)

Electrostatic precipitators and fluidized bed combustion both remove pollutants from the air by turning them into solids, but then the solid waste must be disposed of. The fly ash from electrostatic precipitators contains heavy metals. The sludge from fluidized bed combustion is many times heavier than if the pollutant were a gas.

67. (A)

The first step in a demographic transition is a decrease in the death rate. This may actually result in an increase in population growth for a time, until the birth rate declines.

68. (A)

The epicenter of an earthquake is actually on the surface of the Earth or ocean, just above the focus—or origin—of the quake.

69. (C)

The convergence of two oceanic plates creates a subduction zone—with islands that are formed above the edge of the overlapping plate. Indonesia, Japan, and the Aleutians are examples.

70. **(B)**

Under most typical conditions, it takes about a year to produce one millimeter of topsoil. With little biological abundance and lack of moisture, it would take much longer.

71. **(B)**

Cyanobacteria, or blue-green algae, get the credit for giving us our first oxygen-rich atmosphere. However, trees now help out significantly, particularly rainforests.

72. **(D)**

Mortality is a concept that defines how many people die in a population; but it is also a specific measurement that refers to the total number of people who die in one year per 1,000 people in the population. For example, if a population has 1 million people and a mortality of 20, that means that 40,000 people will die in a two-year period.

73. **(C)**

Subtracting the crude death rate from the crude birth rate means that the population grows by 50 per thousand, or 5 per 100, or 5%.

74. **(E)**

Using the Rule of 70, where 70 = percent growth × doubling time, then the doubling time is closest to 20 years.

75. **(E)**

The Kyoto Protocol addresses gases that contribute to climate change, or greenhouse gases.

76. **(D)**

The national forests are part of the National Forest Service, which is administered by the Department of Agriculture. National forests most closely mirror the philosophy of Gifford Pinchot, who urged the government to set aside lands to benefit future generations.

77. **(D)**

The riparian zone includes the bank of the river, and those areas that are sometimes underwater and sometimes not underwater.

78. **(B)**

If each step is 80% efficient, then it allows 80% of the energy to be used at the next step. The amount of energy that makes it through all three steps is $0.8 \times 0.8 \times 0.8 = 0.51$, or 51% of the original amount of energy.

79. **(A)**

Oil gets the award as the largest supplier of energy among all developed countries, but gas is catching up. However, our coal reserves are more extensive. We think that we have enough coal for about 400 years. However, coal is the dirtiest of fuels, and many communities and utilities are shutting down coal plants. Natural gas gives the cleanest burn, but is not a renewable resource, like solar energy.

80. **(B)**

$$\left(\frac{23}{1000} - \frac{11}{1000}\right) \times 100 = 1.2\% \text{ annual growth rate}$$

81. **(A)**

A pH of 4 is a strong acidic. pH is a measure of acidity. Because pH is a logarithmic scale, each pH level corresponds to a tenfold increase in acid concentration. A pH of 4 corresponds to a hydrogen ion (acid) concentration of 10^{-4} moles of acid per liter of solution. Any pH less than 7.0 is acidic.

82. **(C)**

Metamorphic rock is created when either sedimentary rock or igneous rock is compressed under great pressure and/or heated, without melting. For example, marble is the metamorphic counterpart to limestone, a sedimentary rock, after it has been put under tremendous pressure.

83. **(A)**

Nitrification is the process of oxidizing—or adding oxygen to—nitrogen. Denitrification is the process of removing oxygen from nitrogen. Oxygen must be removed from the nitrate ion in order to become the nitrite ion.

84. **(D)**

The zone of aeration is the portion of the ground through which water percolates as it leaches through the ground on its way to an unconfined aquifer.

85. **(B)**

Fifty percent of the radioactive isotope decomposes after only eight days. That means that after 32 days, or 4 half-lives, the amount will be cut in half four times, leaving only about 6.25%.

86. **(C)**

The Rule of 72 works for this one: 72 = annual growth rate × doubling time. 72 divided by the doubling time (7) equals the annual growth rate as a percentage (about 10%).

87. **(A)**

K-strategists do not generally overshoot carrying capacity. They are more sensitive to environmental resistance and adjust population growth before the carrying capacity is reached. This results in an S-curve on a growth curve graph, rather than a J-curve, which tends to belong to an r-strategist population.

88. **(D)**

Sediment is by far the most prevalent form of water pollution, or spoiling.

89. **(A)**

The wattage of a bulb is the rate at which it uses power. The amount of energy used is the wattage times the time. The bulb uses 200 W for 4 hours, for a total energy consumption of 800 Wh. This is then converted into 0.8 kWh because there are 1,000 W in 1.0 kW.

90. **(C)**

100 grams of water × increase in temp of 20°C × specific heat of water = 8,400 J

91. **(C)**

For the 100 W bulb, 20% of its energy will be given off as heat. That means that each hour, 20 W of energy is given off; or 2,000 W, or 2 kW, over a 100 h period of time.

92. **(E)**

To convert from ppb to ppm, divide by 1,000. There are 1,000 ppb in 1.0 ppm; or 1,000 ppb = 1.0 ppm. The answer, therefore, is 75/1,000, or 0.075 ppm.

93. **(E)**

Integrated pest management involves an integrated approach of many pest-fighting techniques, such as chemicals, natural predators, male sterilization, and using pheromones to lure pests into traps. This technique usually uses fewer chemical pesticides.

94. **(E)**

Unfortunately, all of the items mentioned on this list are consequences of raising beef cattle on feedlots. While it takes less real estate to raise beef cattle this way, it is a chemical-intensive industry that has many adverse environmental effects.

95. **(A)**

Don't be fooled by the word *organic*. In this sense, it means that it involves carbon in the molecule. It takes many centuries for organic material—such as proteins and carbohydrates—to be pressed into peat, then coal. Therefore, it is not considered a renewable process within a time frame that does our species any good.

96. **(C)**

The tilt of the Earth's axis most contributes to seasons. When a particular hemisphere is tilted toward the sun, it experiences summer; when that hemisphere is tilted away from the sun, it experiences winter. Answer (A) would suggest that the entire Earth experiences winter and summer at the same time—which it does not. Answer (B) describes an eclipse, not a contribution to climate. Answer (D) describes night and day. Answer (E) may affect how warm it is during a particular season, but it does not determine the time of the seasons.

97. **(C)**

The carbon dioxide given off by respiration reacts with atmospheric water vapor to form carbonic acid, which is weakly acidic.

98. **(D)**

Soap is not a way to kill waterborne pathogens, but UV light and ozone do kill pathogens and are frequently used in water treatment plants.

99. **(D)**

Ozone in the stratosphere absorbs UV rays. That is important to life on Earth because UV rays are mutagenic and can burn living tissue—both features would make it difficult to have life on Earth.

100. **(B)**

Pronatalist pressure refers to the forces acting on a family that encourage them to have children. Only (B) represents a birth reduction pressure. All the other answers are examples of pronatalist pressures.

Section II

1. (a) (2 points possible)

    +1 for mentioning one of the following acts; +1 for identifying the effect:

    - Endangered Species Act: protects endangered species and their habitats
    - Species Conservation Act: bans importation of endangered species, purchases endangered species' habitat

    Other laws enforced by this agency include: (+.5 point if mentioned, but no explanation offered)

    - Fish and Wildlife Conservation Act
    - Fish and Wildlife Act
    - Fish and Wildlife Improvement Act

(b) (1 point possible) land exchange mitigation

(c) (1 point possible)

Full credit should be given for a detailed description of full-cost pricing or internalizing external costs. This essay is an example:

    While some might be concerned that increasing the price of each home results in excluding low-income families (social injustice), the most important point is that the new housing developments are approaching their true cost. Normally, an external cost of a house that homeowners rarely bear is the cost of a loss of biodiversity and habitat. However, with this scheme, the external cost of lost habitat becomes an internal cost that is paid by the person who is gaining the privilege of displacing rare species.

(d) (3 points possible; +1 for any one of the following answers, to a maximum of 3 points)

- Species uses a logistic growth strategy
- Long life span
- Large body
- Top carnivore
- Low reproductive rate
- Requires a large amount of land for roaming
- Migratory animal has one large biome endangered
- Narrowly defined niche
- Experienced genetic assimilation by exotic species
- Outcompeted by more dominant species for resources
- Particularly sensitive to pollution, or change in climate, or some other environmental resistance

(e) (3 points possible; +1 each for any of these answers, to a maximum of 3 points; +.5 point if reason is given without explanation)

- Water pollution because the protected real estate still obtains water from outside the protected area, such as from runoff or groundwater.
- Air pollution because the protected real estate still obtains air from outside the protected area.
- Segmentation or genetic isolation because the organism may not have a large enough choice of breeding partners in a preserve in order to sustain a minimum viable population.

2. (a) (2 points; +1 for setup and units, +1 for correct answer)

$$3{,}400{,}000 \text{ MWh} \times \frac{3{,}400{,}000 \text{ BTU}}{1.0 \text{ MWh}} = 4.4 \times 10^7 \text{ BTU}$$

(b) (2 points; +1 for setup and units, +1 for correct answer)

$$4.4 \times 10^{11} \text{ BTU} \times \frac{1 \text{ lb coal}}{5{,}000 \text{ BTU}} = 8.8 \text{ million lb coal}$$

(c) (2 points; +1 for setup and units, +1 for correct answer)

$$\frac{0.64}{0.36} \times 4.4 \times 10^{11} \text{ BTU} = 7.8 \times 10^{11} \text{ BTU}$$

(d) (i) (1 point) Warming the river results in thermal pollution, which decreases the ability of the river to dissolve oxygen. Thermal pollution from electrical power production competes with other utilities, such as municipal sewage treatment plants, for the oxygen-dissolving capacity of the river.

(ii) (2 points; +1 for writing the specific heat capacity equation, +1 for the correct answer)

Heat absorbed = mass of water × change in temperature × specific heat capacity

$$\frac{7.8 \times 10^{11} \text{ BTU} \times 0.8 \times {}^\circ\text{F} \times \text{lb}}{5 {}^\circ\text{F} \times \text{BTU}} = 1.2 \times 10^{11} \text{ water per day}$$

3. (a) (3 points maximum; .5 for Point A as the healthiest point; +.5 for each of the following)
    - DO is higher
    - pH is in physiologic range
    - Low nitrate level suggests that water is free of nutrient pollution
    - Higher biodiversity
    - Higher abundance

(b) (1 point possible; +.5 point for each of the following)
    - Abundance
    - Diversity

(c) (2 points possible; +1 point for each of the following)
    - Dissolved solids are measured with a conductivity meter

- Undissolved solids are measured with a Secchi disk and this condition is related to water turbidity

(d) (2 points total)

The high nitrate level suggests that nutrient pollution caused an algal bloom, and then the decomposer bloom depleted the oxygen (+1 point). This is supported by a low DO, and a drop in diversity and abundance. The higher temperature suggests that there is also thermal pollution, which amplifies the effect of nutrient pollution by decreasing the solubility of oxygen in the water (+1 point).

4. (a) (2 points possible; +.5 for each of the following for a maximum of 2 points)
   - Passive design; any one example, such as extended eaves, deciduous trees, trombe wall, floor heating, passive convection, counts for points.
   - Active solar heating, such as direct heat of water by solar collectors
   - Photovoltaic system
   - Geothermal system
   - Wind generator
   - Energy savings; any number of different examples may be used, such as energy-saving appliances, CFLs, etc.

(b) (1 point possible)

Ammonia from human waste is digested by nitrifying bacteria and converted into nitrites, then into nitrates, which are dispersed through the soil and used as nutrients by plant roots.

(c) (3 points possible; +1 for each of the following, +.5 each if no reason given)
   - VOCs from carpets, solvents, paints, adhesives, new material. Solution: use natural fibers, wood floors, eco-friendly paints.
   - Smoking. Solution: don't smoke or allow people to smoke in your house.
   - Radon. Solution: ventilate basement or crawl space.
   - Carbon monoxide. Solution: Do not use combustion appliances; install a CO alarm in your house.
   - CFCs from dry-cleaned clothes. Solution: Use clothing that does not require dry cleaning or choose an alternative like pressurized carbon dioxide or so-called wet cleaning.
   - Fiberglass fibers from insulation. Solution: Use insulation from recycled materials or cellulose.

(d) (1 point possible)

Answers may come from the following list:

- Sight selection: close to community and transportation, reuse of brownfield, maintain open space
- Water efficiency: use reduction with efficient appliances, xeriscaping
- Use of eco-friendly materials: material reuse, materials from nearby sources
- Compatibility with regional priorities: locally recycled products, renovating blighted areas, etc.

(e) (3 points possible; +.5 for each of the following)

- Heat home through geothermal or solar (passive or active), rather than burning fossil fuels.
- Regularly service furnace to be sure it is burning efficiently.
- Use wood from sustainable forests.
- Position home close to public transportation and/or stores.
- Use materials from nearby sources, or recycled materials, to minimize transport to the building site.
- Use photovoltaic energy on site to charge an electric car.
- Use less plastic in construction; use renewable materials (such as wood from sustainable forests) instead.
- Plant vegetation, including a personal garden, to both sequester carbon and reduce trips to the grocery store.
- Use LED and other energy-efficient lighting, which uses less fossil fuels.
- Use insulation to minimize heat loss, thereby using less fossil fuels.

Glossary

abiotic factor condition in the physical, nonliving environment to which organisms are subjected, such as temperature, salinity, light intensity, or pH.

abundance with regard to living individuals, a measure of the total number of organisms of each species in a particular ecosystem.

abyss a benthic marine biome existing at depths greater than 2,000 meters; sometimes also applied to the zone in lakes below the photic zone.

acid a substance that releases hydrogen ions (H^+) in water solution, or accepts electron pairs in chemical reactions; a solution with a pH below 7.0.

acid rain rain or any other form of precipitation that has become acidic (below pH 7.0) by absorbing air pollutants, especially oxides of nitrogen or sulfur.

acute toxicity a large dose of a toxic substance that inflicts immediate harm on an organism (compare to chronic toxicity).

adaptation a genetically controlled change in structure, function, or other characteristic that may make an organism better able to survive in its environment.

aerosol effect blocking the sun's rays from particulate matter, which works to decrease the amount of solar energy that reaches the Earth.

age-structure diagram histograms that reveal the distribution of people at different ages within a population.

air stripping a technique that removes volatile pollutants from soil using hot gas at high pressure.

allergen a biotic or an abiotic microscopic entity that initiates an allergic response.

alpha decay ionizing radiation that emits a particle with two protons and two neutrons, or a helium nucleus.

amino acid an organic molecule that includes one or more amino (NH_2) and acid (COOH) groups; 20 amino acids are essential for human survival, which are linked together during protein synthesis to become proteins.

ammonification The process by which decomposers convert animal and plant protein into ammonia.

anthropogenic caused by humans.

aphotic zone deeper parts of aquatic habitats where light does not penetrate.

aquifer geologic formation through which water can percolate and accumulate; porous rock saturated with water.

arithmetic population growth linear growth, one that increases by a constant amount over time.

assimilation the process by which plants absorb ammonia and nitrate ions to build large, essential molecules, such as proteins and DNA.

asthenosphere the less rigid, flexible layer of the Earth's interior, between 70 to 150 and 200 to 360 kilometers below the surface.

atom the basic unit of matter; the smallest complete unit of an element, consisting of protons, neutrons (usually), and electrons.

bacterial resistance the phenomenon where disease-producing bacteria are no longer sensitive to the antibiotics that have been used to destroy them in the past.

base a substance that releases a hydroxide ion (OH^{-1}) in water solution, or gives up an electron pair in a chemical reaction; also, a nitrogen-containing chemical compound, such as a purine or pyrimidine, found in nucleic acids.

benthic pertaining to the bottom zone, or bed, of lakes, seas, and oceans.

beta decay ionizing radiation consisting of electrons ejected by an unstable atomic nucleus.

bioaccumulation the process by which substances, especially fat-soluble toxins, build up in the tissues of living organisms.

bioassay quantitative measurement in the laboratory on the effects of a dose of a substance on an organism or part of an organism.

biodiversity the number of available species in an ecosystem (species biodiversity); the amount of genetic variation present in a population's gene pool (genetic biodiversity); or the range of habitat available to support life (ecosystem biodiversity).

biological control a method of reducing the population of an undesirable species by introducing a natural enemy of that species.

biological oxygen demand a measure of the amount of oxygen required to oxidize matter in water samples; an estimate of the load of oxygen-consuming organisms in an aquatic ecosystem.

biomagnification increase in the concentration of toxins in successively higher trophic levels in a food chain or food web.

biomass mass of living and formerly living tissue.

biome ecosystem; consists of a large community of different populations combined with the abiotic factors surrounding those populations.

bioremediation the use of living organisms to remove hazardous or toxic pollutants from the environment.

biotic factors influences on the environment from living and formerly living material.

biotic potential the maximum reproductive rate of an organism, given unlimited resources and ideal environmental conditions, with no environmental resistance.

BLM abbreviation for the Bureau of Land Management, a government agency within the Department of the Interior.

boreal *see* taiga forest

bottleneck effect a limitation on genetic biodiversity because a cataclysmic event has left only a few survivors, so that the gene pool of the subsequent population is limited by the available genes that the survivors transmit to offspring. Also called *genetic bottleneck*.

British Thermal Unit (BTU) a unit of heat energy where 1.0 BTU equals the amount of heat needed to raise 1.0 pound of water by 1.0°F.

Bureau of Land Management *see* BLM

calorie amount of heat required to raise one gram of water by 1.0°C.

carbon cycle the chain or cycle of reactions by which carbon is circulated through the environment, especially via the process of photosynthesis and respiration.

carbon sequestration the storage of carbon in various forms within biomes, such as forests, or as methane frozen within the tundra floor.

carcinogen a substance or agent capable of causing cancer.

carrying capacity the number of individuals in a population that can be supported by the resources in that habitat.

chaparral a dry terrestrial biome characterized by short, wet winters, and long, dry summers; flora includes thickets of low evergreen oaks or dense underbrush; for example, the southwestern United States.

chemical synergy the impact of two toxins or drugs having a greater effect upon simultaneous exposure than the sum of the effect of the two toxins or drugs would have by themselves.

clear-cutting logging practice where every tree, regardless of species or size, is harvested and removed.

climax community a relatively stable combination of populations that result when succession of an ecosystem reaches a point that resists further change; the final stage of ecological succession.

cold front moving boundary of cold air that displaces warmer air.

commensalism a relationship between organisms in which one individual benefits from living close to, or on, another organism, and the "host" organism remains unaffected.

community a collection or group of interdependent populations of different species whose niches overlap by geographical location and time.

competition process by which more than one species, or individuals of the same species, strive to occupy the same niche and survive.

compound a chemical substance composed of two or more elements, joined by chemical bonds, and existing in a fixed proportion.

conduction the conveyance of energy, such as heat, sound, or electrical, through a material as a result of collisions at the atomic level.

coniferous forest a biome of cone-bearing evergreen trees with needle-like leaves; found in temperate zones and colder forests, such as a taiga.

conservation the planned management and wise use of natural resources for present and future generations; the use, protection, and improvement of natural resources according to principles that will ensure their highest economic or social benefit.

constructive boundary an elongated region in which magma is pushed up from the asthenospheric mantle, producing new crust and pushing apart tectonic plates. Also called a *divergent boundary*.

consumer a heterotrophic organism that obtains nourishment through the consumption of other organisms.

continental drift the plate tectonic process by which land masses slowly and continuously move positions on the globe, creating new land masses by colliding or splitting apart.

convection the transfer and transport of heat energy by the movement of a gas or fluid; the primary mechanism of energy transfer in the atmosphere, oceans, and within the Earth.

conventional pollutants any of seven major pollutants as designated by the Clean Air Act (sulfur dioxide, carbon monoxide, particulates, hydrocarbons, nitrogen oxides, photochemical oxidants, and lead); pollutants that make up the largest volume of air quality degradation and pose the most serious threat to human health. Also called *criteria pollutants*.

convergent evolution the evolution of two different groups of organisms or species that come to resemble one another, or occupy a similar niche in different ecosystems.

convergent plate boundary the elongated region of collision between tectonic plates.

coral reef a type of marine ecosystem based on a mutualistic relationship between photosynthetic algae and a marine colonial coelenterate that secretes a calcareous skeleton, forming elaborate structures in shallow ocean waters or along shelves in warm, shallow, tropical seas.

Coriolis effect the deflecting force acting on a body in motion due to the combination of convection and the rotation of the Earth. As a result, the moving body rotates toward the right in the Northern Hemisphere and toward the left in the Southern Hemisphere.

cost-benefit analysis economic technique used to evaluate large public projects where the desirability of a proposed course of action is estimated by listing and comparing advantages and disadvantages.

criteria pollutants any of seven major pollutants as designated by the Clean Air Act (sulfur dioxide, carbon monoxide, particulates, hydrocarbons, nitrogen oxides, photochemical oxidants, and lead); pollutants that make up the largest volume of air quality degradation and pose the most serious threat to human health. Also called *conventional pollutants*.

crude birth rate quantitative measurement for fertility, expressed as the number of births per thousand people in a population.

crude death rate quantitative measurement for mortality, expressed as the number of deaths per thousand people in a population.

crust the solid outer shell of the Earth, varying in thickness from 6 km (under oceans) to as much as 70 km in continental regions.

decomposer an organism that obtains nutrients from feeding upon dead organisms, breaking them down into simpler substances, which become nutrients for other organisms.

demographic transition changes in a population that lead to a combination of low birth and death rates.

denitrification stage in the nitrogen cycle where bacteria turns oxidized nitrogen (nitrates and nitrite ions) into nitrogen gas.

desert a very arid terrestrial biome characterized by low moisture—usually less than 25 cm of rainfall annually—and little vegetation; seasonal temperatures fluctuate widely.

desertification the process that transforms semiarid ecosystems into deserts, usually through a loss of primary productivity, moisture from transpiration, soil erosion, or salinization.

destructive plate boundary elongated region of tectonic plate collision where one or more plates is destroyed. *See* subduction zone.

dioxin a compound that is extremely toxic to plants and animals, very persistent, and capable of causing chromosome damage.

dissolved ions charged particles whose ionic attractions have been overcome in an aqueous environment so that the dissociated ions have dispersed throughout the solution.

dissolved oxygen (DO) a measurement, usually in parts per million (ppm), of the amount of oxygen gas dissolved in an aquatic environment.

divergent evolution the situation in which comparative structures on different organisms evolve to perform different functions, or have increasingly diverse morphology.

diversity a measure of the number of species in an ecosystem, number of ecosystems in the world, or the number of genetic variations within a species.

dominant species a species that exerts an overriding influence in determining the characteristics of a community.

dose the amount of a chemical substance, microorganism, or energy that is taken into or absorbed by a body. Dose is determined by both amount and duration of exposure.

dose-response curve a graph that depicts the relationship between the degree of exposure to a substance (dose) and the observed biological effect or response.

doubling time the amount of time that it takes for a population to double, approximated by dividing the percent growth rate into 70.

ectoparasite a parasite living on the outside of a host, such as a tick or a flea.

edge effect a change in biotic and/or abiotic factors at the boundary between two ecosystems; often involves greater biodiversity than that of the individual ecosystems.

El Niño Southern Oscillation (ENSO) a large-scale change in normal weather patterns and ocean currents of the Pacific Basin. Nutrient-rich upwelling currents along the coast of South America are stalled, limiting productivity and causing fisheries to fail. An El Niño event normally is accompanied by droughts in Australia and Southeast Asia, and heavy rain and snow in western North America.

emergent disease diseases that are either new to a population or are reappearing after having been absent for some time.

emigration the movement of members of a population to another location.

endangered species a species that is in danger of becoming extinct.

energy a quality of light and matter that is transferable from one place to another through conduction, convection, or radiation. Units are $kg \times m^2/s^2$.

energy efficiency a measure of energy produced compared to the energy consumed.

energy pyramid a representation of the loss of energy at each step in a food pyramid.

entropy the internal energy of a system that cannot be converted to mechanical work; a measure of the disorder of a system, where some energy is lost to the random disbursement of waste heat; the loss of energy at each trophic level of an ecosystem.

environment the sum total of all the external conditions within which an organism lives.

Environmental Impact Statement an assessment of the environmental consequences of a course of action, used as part of a decision-making process and required by provisions in the National Environmental Policy Act of 1970. It is required of any major project that a U.S. federal agency intends to undertake.

environmental resistance all the limiting factors, both biotic and abiotic, that tend to suppress population growth and set the overall carrying capacity of an ecosystem for any given population.

epilimnion the upper layer of warm oxygen-rich water in a thermally stratified lake or reservoir, extending from the surface to a depth of 5 to 10 meters; lies above the thermocline.

essential amino acids the eight amino acids that cannot be synthesized by the human body and must be consumed.

estuary the interface between river and coastal ecosystems, often the mouth of a river, or a semi-enclosed coastal environment that has a high influx of freshwater and salinity.

euphotic zone surface layer of water where phytoplankton can live; the depth of light penetration at which photosynthesis balances respiration.

eutrophication the situation in which an excess of nutrients—usually nitrogen- or phosphate-based—are introduced into an aquatic habitat, producing dramatic growth of phytoplankton. As the algae die, decomposers use up the oxygen in the water, which decreases the amount of dissolved oxygen in the water.

evolution the process by which the gene pools of species change as a result of variable survival of individuals who express different genetic traits in different environmental conditions.

exosphere atmospheric layer 300-6,000 miles above the surface of the Earth, forming the transition to interstellar space, and composed primarily of hydrogen and helium.

exponential population growth a condition where each new generation of individuals produces more potential reproducers than the previous generation. Growth is slow at first and then rapidly accelerates beyond the carrying capacity of the habitat. Also called *geometric growth* and *logarithmic growth*.

external cost the negative effects of production or consumption for which no compensation is paid, such as pollution, where the actions of the polluter have an adverse effect on people or the environment, but the polluter does not experience the cost of that pollution.

extinction the disappearance of a species or gene, globally or locally.

famine massive, acute incidence of undernourishment that is usually catalyzed by political, economic, and/or environmental upheaval.

fecundity the ability to reproduce offspring; the rate at which an individual reproduces.

fertility a measure of the actual number of offspring produced, expressed statistically as the crude birth rate.

First Law of Thermodynamics law of physics stating that energy is neither created nor destroyed; it can only change form.

flocculation the process in sewage treatment that uses aluminum sulfate and other chemicals to cause very small particles in sewage sludge to come together to form larger accumulations, which can then be removed by filtration.

food chain an abstract representation of the sequence of organisms that traces the movement of biomass or energy, through trophic levels, from producer to consumer.

food web an abstract representation of interconnected food chains within a community.

fossil fuels the fuels that result when geological processes of great pressure, heat, and time act on organic remains.

fracking hydraulic fracturing; the process by which water, sand, and chemicals are pumped into wells so that additional oil or gas is extracted from the ground.

fragmentation when human development or natural catastrophe has turned a large contiguous ecosystem into a patchwork of subpopulations that are unable to interbreed, or they have a limited roaming region.

fugitive emissions air pollutant category, as defined by the Clean Air Act of 1970; pollution that does not come from a point source, such as a smokestack, but from nonlocalized sources, such as automobile emissions.

fusion nuclear reaction that combines smaller atoms into larger atoms with the release of great amounts of energy.

gene a unit of genetic inheritance; the portion of DNA containing information needed to produce a specific protein or RNA molecule.

gene pool the collective group of genetic traits (expressed and nonexpressed) that exist in all of the chromosomes of all the individuals in a population.

genetic assimilation the blending of the genetic traits of a subspecies into the gene pool of a closely related, more hardy species, ultimately leading to the extinction of the subspecies.

genetic biodiversity the breadth of different genetic traits (both expressed and nonexpressed) in a population's gene pool, or within the genes of a single individual.

genetic bottleneck *see* bottleneck effect.

genetic isolation an event occurring when a small number of individuals has been isolated from the rest of the population; as a result, their gene pool becomes limited to the genes available in the few breeding individuals, ultimately leading to genetic incompatibility with the original species.

genetically modified organism an organism with a genetic constitution that has been altered in some manner by human intervention.

geothermal energy energy obtained from the Earth either through thermal features, or by using the constant temperature of the Earth in combination with a heat pump to heat or cool a building.

global warming phenomenon in which the average annual temperature of the Earth increases. *See* greenhouse effect.

grasshopper effect phenomenon where a trade wind and the Coriolis effect eventually create the accumulation of toxins in the polar regions.

green revolution cultural and scientific movement that created high-yielding grains that would hopefully solve the world's hunger problem, but which is fraught with all the problems of large-scale monoculture.

greenhouse effect the process by which solar heat radiation is trapped by gases in the atmosphere; the warming of Earth's climate by the concentration of carbon dioxide and other gas pollutants in the atmosphere.

groundwater water that is stored within porous rocks underground, as with an aquifer.

growth rate the sum of the increases to a population due to immigration and births minus those individuals who have died or emigrated.

habitat the place in the environment where a particular organism, or community of organisms, lives.

hadal zone the deepest marine trenches, usually greater than 6,000 feet in depth.

homeostasis a physical state of health with a balanced operation of all body systems; the maintenance of constant internal conditions despite a varying external environment.

hydrologic cycle the movement of water through the atmosphere, bodies of liquid water, in the Earth, and through ecosystems; includes the passage through living organisms.

hypolimnion the most dense and deepest layer of water in a lake.

igneous rock rocks formed from the solidification of magma.

ignitability the ability of a chemical to ignite into a combustion reaction; a quality of hazardous chemicals.

indicator species a species that serves as an early warning that some parameter of an ecosystem has changed.

industrial revolution a period in human history beginning in England in the eighteenth century, when agricultural technology, the availability of coal, and steel-making technology coincided to produce industrialized economies.

industrialized monoculture an agricultural practice where large tracts of land are planted with a single crop, and the same maintenance techniques are implemented over a broad region.

integrated pest management procedures that involve a combination of pest control strategies—chemical and nonchemical—that are specific to the pest, crop, and location.

internal cost the cost of production that is directly borne by the producer or consumer of a product.

interspecific competition niche overlap and subsequent competition between individuals of different species.

intertidal zone the region between low and high tides where organisms are subjected to the forces of moving water and waves, and the periodic exposure to open air and saltwater immersion.

intraspecific competition competition for resources or mating partners among members of the same species.

ion atoms or combinations of atoms that have an unbalanced electrical charge; the dissociated parts of a molecule carrying either a positive or negative charge.

ionosphere the region of the atmosphere that extends 50 to 300 miles above the Earth's surface and is made of multiple layers dominated by electrically charged, or ionized, atoms.

J-curve a type of population growth curve that demonstrates the maximum biotic potential of a population; exponential growth; named for the shape of the curve.

joule metric unity of energy; the work done when a force of 1 Newton displaces an object by 1 meter.

k-strategy the reproductive growth strategy of populations that tend to respond more quickly to environmental resistance and experience a sigmoidal, or s-curve, growth curve.

kelp forest a marine habitat characterized by dense growth of kelp, a marine algae; found in shallow areas of temperate seas and oceans.

keystone species species that is critical to the ecological structure of an entire biological community.

kinetic energy energy that an object has by virtue of the velocity at which it travels or moves.

Kyoto Protocol international agreement among countries to restrain the production of gases that contribute to global warming.

Law of Conservation of Mass *see* First Law of Thermodynamics

LD_{50} the concentration of a substance that causes death in half of a population that is exposed to that substance.

LEED (Leadership in Energy and Environmental Design) set of building criteria established by the United States Green Building Council that minimizes a building's overall pollution and use of energy.

life expectancy the number of years an individual is statistically predicted to survive.

lifespan the longest length of life reached by a given species.

limnetic zone the portion of open water in a lake or other deep freshwater sea, below the euphotic zone, that cannot sustain rooted vegetation.

lithosphere the outer solid part of the Earth, including the crust and the uppermost mantle that is about 100 km thick.

littoral pertaining to the shore or beach environment in marine ecosystems, or to the shallow area of a lake that can sustain rooted plants in freshwater lakes.

logarithmic growth *see* exponential population growth

magma molten volcanic rock.

malnourishment the condition in which a living organisms gets enough calories to survive but does not get the right amounts of vitamins and nutrients to stay healthy.

mangrove swamp coastal ecosystem that has ocean water flowing around the roots of mangrove trees, which serve as keystone species and habitats for numerous other species.

mantle the layer of Earth lying between the crust and the core extending from about 10 km under the oceans and 30 km under continents, to about 2,800 km from the surface of the Earth.

mesolimnion the middle-layer temperature zone of a thermally stratified lake or other deep body of water; between the upper warm epilimnion and the lower cold hypolimnion.

mesosphere atmospheric layer 30 to 50 miles above the Earth's surface where temperature decreases with altitude; contains ice crystal clouds.

metamorphic rock sedimentary and igneous rock that has been altered in texture and composition as a result of being exposed to great heat or pressure, without passing through a molten stage.

Minimata disease a disease of the central nervous system caused by mercury poisoning; named after a bay and town in Japan whose inhabitants experienced such poisoning by consuming contaminated fish and shellfish.

mitigation in the context of land use, the creation of an ecosystem of comparable health in exchange for damage done by the development of another area nearby.

Montreal Protocol an international agreement among several countries to reduce the production of gases that contribute to ozone depletion.

morbidity a measure of the level of illness in a population.

mortality a measure of the actual number of individuals, per thousand in the population each year, who die.

mutagen a chemical that causes genetic change.

mutation a change in base sequences within a cell's chromosome, changing the DNA and usually causing damage to the organism.

mutualism a type of symbiosis whereby both species in the relationship benefit.

natality a measure of the production of new individuals in a population.

national forests government-owned forests, managed by the Department of Agriculture, that attempt to create "the greatest good for the greatest number of people."

national parks government-owned historic or natural treasures, managed by the Department of the Interior, so that succeeding generations may learn from and be inspired by the protected area.

natural population growth rate population growth as measured only by the difference between birth and death rates.

natural selection the selective survival of individuals within a species based on the ability of their unique combination of traits to survive in the environment in which they live.

Neolithic Revolution a period in human history that represents the shift from a nomadic hunting/gathering lifestyle to an agricultural lifestyle, where populations settle in areas close to food, water, and rich soil.

neritic zone the marine ecosystem that consists of relatively shallow water extending from the littoral zone to the edge of the continental shelves.

niche the combination of habitat and the ecological role an organism plays in an ecosystem.

nitrification the process by which ammonia is converted into nitrite; one step in the nitrogen cycle.

nitrogen cycle the chain or cycle of events by which nitrogen is circulated through the environment and living organisms.

nitrogen fixation the process by which atmospheric nitrogen is converted into ammonia; one step in the nitrogen cycle.

nonrenewable resource a natural resource present on Earth in a fixed quantity that cannot be replaced during human time scales and can be effectively used only once, such as fossil fuels.

oligotrophic aquatic habitats that are poor in plant nutrients and tend to have little biological productivity.

oxygen-demanding wastes pollutants, such as nitrates and phosphates, that result in a surge of algal growth, followed by an increase in decomposing organisms, which use up the oxygen in a stream or lake, thereby making it less habitable to other organisms.

ozone a molecule that contains three atoms of oxygen. It is a caustic pollutant to humans, but is a critical protector when it exists in the stratosphere, where it absorbs harmful ultraviolet light that would damage living tissue.

ozone hole a thinning of the upper atmosphere ozone layer due to a reaction with human-made chlorinated hydrocarbons, thereby reducing the amount of protection from ultraviolet light that is harmful to living tissue

parasitism an association of organisms where one organism derives all of its nutrients from a host.

particulate matter a type of air pollution that is composed of small individual solids, such as smoke.

pathogen disease-causing microscopic organisms or viruses.

pelagic pertaining to the open sea; an aquatic biome characterized by the upper, open waters of a lake, sea, or ocean, where actively swimming creatures and suspended plankton exist, and no land is available.

pH the scale of acidity that ranges from 1 to 14, where values less than 7 are acidic, and values greater than 7 are alkaline; the logarithm of the concentration of hydrogen ions in an aqueous solution.

phosphate cycle the chain or cycle of events by which phosphate ions circulate through the environment and living organisms.

photic zone the upper layer of a lake, sea, or ocean where there is enough light for photosynthesis to take place.

photochemical oxidants secondary air pollutants that are synthesized with the aid of solar energy, such as smog and ground-level ozone.

photosynthesis the cellular process in green plants and some bacteria by which carbon dioxide and water are combined in the presence of light energy to produce simple carbohydrates and oxygen.

photovoltaic an apparatus that converts energetic photons from sunlight directly into electrical energy.

phytoplankton microscopic floating plants in aquatic ecosystems; the basic organism in aquatic food chains.

pioneer species the first of a successive series of species invading a newly opened habitat.

plate tectonics the scientific theory stating that the Earth's crust is made up of several large, rigid plates that move by convective forces below the crusts.

plates rigid slabs of Earth's crust and upper mantle that make up the lithosphere.

point source pollutant pollution discharged from a specific point, or location, such as a pipe.

population an interbreeding group of organisms of the same species that lives in the same general area over a given period of time.

precautionary principle a principle that places the burden of proof on the person or group of people who are changing an ecosystem, suggesting that greater potential for damage to the ecosystem should demand greater certainty about how to prevent the damage.

precipitation moisture that reaches the Earth's surface from the atmosphere, including rain, mist, dew, sleet, snow, and hail.

preservation the act of reserving, protecting, or safeguarding a portion of the natural environment from unnatural disturbance. Preservation suggests that natural resources will be left undisturbed.

primary pollutants air pollutant category as defined in the Clean Air Act; pollutants harmful to humans in the form in which they are initially released.

primary productivity the amount of biomass produced by photosynthetic organisms; a measure of how much solar energy is converted into chemical energy per area per unit of time.

primary succession initial sequence of communities that develop in a newly exposed habitat; species that colonize a primary succession are called pioneer species.

primary treatment the removal of particulate matter and sludge from sewage by filtration and settling tanks.

producer organism that manufactures nutrients from inorganic sources by a process such as photosynthesis or chemosynthesis.

productivity the amount of biomass that is produced by a community.

protein a collection of amino acids that form into a macromolecule; frequently used in sustaining living physiologies.

rad radiation absorbed dose. The unit of power of radiation absorbed by an organism, expressed as energy per gram of absorbing material; 1 rad equals 10^{-2} joules of energy absorbed.

radiation energy emitted in the form of electromagnetic waves; the process by which an object emits energy, such as the atmosphere and ground surface being heated by the radiant energy from the sun.

radioactivity the spontaneous disintegration of atoms emitting radiant energy, sometimes also including atomic particles.

radon a naturally occurring radioactive gas that can pose a human health hazard upon inhalation; frequent culprit in indoor air pollution.

rainforest terrestrial biome characterized by heavy rainfall; no marked dry season; and rapid, lush growth of vegetation.

reclamation in environmental management, the rehabilitation of a massively scarred, denuded, or otherwise devastated area to a condition that is environmentally useful and socially or politically acceptable.

recycle to use a resource more than once, either in the same or another form.

rem roentgen equivalent for man; a more accurate unit than rad for assessing biological risk from a dose of radiation; one rem equals the number of rads multiplied by a constant that is related to the type of particle causing the radiation.

remediation in environmental management, the use of chemical, biological, or physical methods to remove hazardous or toxic pollutants.

renewable resource a resource that is continually replenished within a human lifetime, such as wood or sunlight.

respiration the cellular process in organisms whereby oxygen and simple carbohydrates are converted into energy-storing phosphate molecules and carbon dioxide is released.

restoration in environmental management or land-use contexts, returning a damaged ecosystem back to its unspoiled, natural condition.

Richter scale a logarithmic scale that depicts the strength of an earthquake; the higher the number, the greater the intensity of the earthquake.

rift valley a place on a continent where upwelling convective forces in the Earth's mantle cause a separation of the crust, forming new plate material.

riparian zone the terrestrial biological zone on and near the bank of a river or stream.

r-strategy the reproductive growth strategy of a population that tends to reproduce close to its biotic potential, exceed its carrying capacity, then die back after using up too many resources.

salinity a measure of the dissolved salts in a body of water or soil.

salinization the destruction of agricultural land by an increase in soil salinity, frequently occurring from overirrigating soil and then having the water evaporate—leaving behind salts that were in the water.

salt marsh *see* estuary

savannah a terrestrial biome characterized by 50 to 150 cm of rainfall annually, similar to temperate grasslands, but with scatter trees and existing in tropical latitudes.

s-curve sigmoid-shaped population growth curve for populations that respond more quickly to environmental resistance; such populations typically have a k-strategy for reproduction.

seafloor spreading the mechanism by which new seafloor crust is created as adjacent tectonic plates move apart at a constructive boundary, and magma is pushed up from the asthenospheric mantle.

Second Law of Thermodynamics law of physics; no change that involves a transfer of energy from one form to another occurs without some portion of that energy being lost as waste heat, called entropy; no change or reaction is 100% efficient.

secondary pollutant air pollutant category as defined in the Clean Air Act; pollutants released in a form that is initially not harmful but become toxic or hazardous after being transformed in the environment, such as acid rain and photochemical smog.

secondary productivity the amount of biomass produced by organisms that eat photosynthetic organisms, indirectly measured by the amount of waste that a consumer eliminates.

secondary succession the progression of communities when a change creates a new habitat, such as when a climax community is severely disturbed or destroyed.

secondary treatment a second stage of sewage treatment where wastewater is held for a longer time in conditions that are favorable for bacterial digestion of carbon and nitrogen wastes.

sedimentary rock rock made of layers of sediments formed from weathering processes that break down other rocks.

selective cutting a forestry practice that harvests a portion of the mature trees with minimal disruption to a habitat.

sigmoid curve an S-shaped curve for population growth over time in which environmental resistance causes the population growth to approach, but not exceed, the carrying capacity.

soil a combination of sand, clay, and organic material that is conducive to plant growth.

soil horizons vertical layers in the soil that contain different materials with different chemical and physical properties.

soil profile characterization of the vertical cross section of different soil horizons.

speciation the creation of new species through the process of evolution by natural selection.

species a group of organisms with similar enough genetic makeup to be able to reproduce and produce viable offspring.

specific heat capacity a measurement of a material's ability to absorb heat energy.

spring turnover in deep freshwater lakes, a process involving the mixing of surface and subsurface lake waters due to seasonal shifts of temperature. In winter, less dense ice rests on more dense water. In spring, lake surface temperatures warm, causing surface water to sink and mix with deeper water. The process reverses in the fall (fall turnover).

stratopause atmospheric layer 30 to 32 miles above the Earth's surface, forming a boundary between the stratosphere and the mesosphere, where temperature is constant.

stratosphere atmospheric layer 13 to 30 miles above the Earth's surface, where temperature increases with altitude because ozone absorbs ultraviolet solar radiation.

subduction zone a destructive plate boundary, or convergence, where one tectonic plate is forced to slide underneath another tectonic plate.

succession the nonseasonal, continual process where niches and the composition of a biological community change over time.

sulfur cycle the chain or cycle of events by which sulfur is circulated through the environment and living organisms.

Superfund popular name of the hazardous waste cleanup fund established by the Comprehensive Environmental Response, Compensation and Liability Act (CERCLA), which finances the removal and remediation of hazardous waste.

survivorship the number of individuals in a population within a given age bracket that continue to stay alive each year.

survivorship curve a graph that plots the log of survival on the y-axis versus the total lifespan on the x-axis.

sustainable development development where consideration is given to the quality of life for future generations indefinitely.

swidden agriculture agricultural practice used by indigenous people in tropical rainforest areas where small plots of land are cleared for farming. After a few crop yields, a new area is cleared and the old area is allowed to regenerate. Also called *Milpa agriculture*.

symbiosis an ecological relationship between two species, usually involving coevolution; major types of symbiosis include parasitism, commensalism, and mutualism.

taiga forest a terrestrial biome characterized by spruces and firs, with a short summer and long, snowy winter; extends across much of North America and Eurasia; also called boreal forest.

temperate rainforest a rainforest biome occurring at higher latitudes where coastal mountains cause high precipitation; contains mosses, ferns, and large conifer trees. An example is the Olympic Peninsula in Washington State.

temperate zone the part of the Earth's surface characterized by temperate climate located between the Arctic Circle and the Tropic of Cancer, or between the Antarctic Circle and the Tropic of Capricorn.

teratogen a substance or agent that can cause birth defects.

tertiary treatment a third stage in wastewater treatment that involves any of several methods to remove chemical and disease-causing organisms.

thermocline temperature gradient; the boundary between layers of water in a thermally stratified lake or other body of water.

thermohaline circulation the convection cycle in aquatic ecosystems created by differences in temperature and salinity.

thermosphere atmospheric layer 52 to 300 miles above the Earth's surface where temperature increases with altitude because of high-energy solar radiation; also includes the ionosphere.

threshold level the dose whereby an organism begins to experience the desired effect of a drug, or the harmful effects from a toxin.

total population growth rate population growth as measured by the difference of the birth rate and the death rate, added to the difference between immigration and emigration rates.

toxic colonialism the situation in which one country that is eager to dispose of toxic waste takes advantage of an underdeveloped country that is so economically desperate that it is willing to trade the long-term health of its population for short-term cash.

toxin a poison formed as a specific secretion by an organism; any substance that can cause disease.

Tragedy of the Commons the title of an essay by Garret Hardin that describes the lack of responsibility that people feel when they draw upon a resource shared by others.

trombe wall a massive wall used in building construction to absorb radiant energy from the sun, to be reradiated within the building over time.

trophic level a step along a food chain; a particular position an organism occupies in an ecosystem determined by the number of energy-transfer steps leading to that level.

tropical rainforest a rainforest biome in the tropical zone characterized by lush vegetative growth and substantial rainfall of about 200 cm annually.

tropopause atmospheric layer 7 to 13 miles above the Earth's surface, forming the boundary between the troposphere and the stratosphere.

tundra a terrestrial biome in the arctic regions characterized by a short growing season and low temperatures.

undernourishment not having enough food to develop or function normally; less than 2,000 to 2,500 calories each day for humans.

upwelling vertical movement of water that brings nutrients from the depths of an aquatic or marine ecosystem to the surface layers.

urbanization the conversion of agricultural, forested, or other undeveloped areas to urban land as a result of natural population growth and movement of people to urban areas.

vector an agent that transfers a pathogen from one organism to another, such as a tick or a mosquito.

volatile organic compounds (VOCs) a class of organic compounds mainly composed of hydrocarbons that are easily vaporized and pose health and environmental hazards.

waste stream the sequence that solid waste undergoes, from discovering its raw materials, mining, transporting, refining, manufacturing, marketing and delivering a product, use, and/or waste.

water compartment a location in the Earth's biosphere that impounds water in liquid, solid, or gaseous form.

water cycle the cycle by which water moves among water compartments in the environment and among living creatures.

weather front massive movement of air created when an air mass of one temperature and pressure collides with an air mass of a different temperature and pressure.

wilderness area a region that is protected from human machines and development for the benefit of species and ecosystem biodiversity and is managed by the Department of the Interior.

zero population growth a demographic situation in which a population is held at stable numbers over an extended period of time.

zooplankton tiny floating marine animals; together with phytoplankton, they compose the plankton life that forms at the surface of aquatic ecosystems.

Index

A

Abiotic factors, 58
Abundance, in species biodiversity, 69
Abyss, 63
Acid rain, 134, 154, 162–163, 169
Acids, 24
Acute toxicity, 182
Adaptation, natural selection, 68
Aerosol effect, 163
Africa, managing population growth, 89
Age-structure diagrams, 85–86
Agriculture, 94–100
    environmental impact of, 99–100
    essential amino acids, vitamins and minerals, 95
    famines, 96
    feeding growing population, 94
    genetically modified organisms (GMOs), 98
    Green Revolution, 98
    irrigation methods, 97
    soil conservation, 97–98
    Swidden agriculture, 104–105
    toxins in food, 96
    undernourished *vs.* malnourished, 95–96
    water use for, 117
A-Horizon, 53
Air
    atmospheric circulation, 44–45
    global air movement, 44–45
    jet stream, 45
    trade winds, 45
Air pollution, 152–165
    acid rain, 162–163
    aerosol effect, 163
    asthma triggers, 162
    carbon oxides, 155
    criteria and noncriteria pollutants, 153
    effects on humans, 161–162
    Greenhouse effect, 163
    indoor, 159–160
    laws relating to, 164–165
    lung irritation, 161
    measuring, 160
    metals and halogens, 157
    nitrogen oxides, 154
    ozone depletion, 163–164
    particulate matter, 156
    photochemical oxidants, 158
    sources of, 152–153
    sulfur oxides, 154–155
    summary of major pollutants, 158
    as urban environmental problem, 109
    volatile organic compounds (VOCs), 155–156
    weather and, 164
Air stripping, 184
Albedo effect, 194
Allergens, 181
Alpha decay, 138
Amino acids, essential, 95
Ammonification, 72
Andes Mountains, 38
Anemia, 95
Anthracite coal, 134
Anthropogenic sources, 152
Antibiotics, in food, 96
Antihistamine, 181
Aphotic zone, 63
Aquaculture, 103
Aquatic biome, 63–64
Aral Sea, 119
Argon, percent in atmosphere, 42
Ariboflaninosis, 95
Army Corps of Engineers, 118
Asbestos, 159
Assimilation, 72
Asthenosphere, 36
Asthma triggers, 162
Atmosphere
    atmospheric circulation, 44–45
    climate, 46–47
    composition of, 42–43
    creation of oxygen in, 40
    layers of, 41–42
    weather, 45–46
Atmospheric layers, 41–42
Atoms, 23
Automobile society, energy use in, 133
Autotrophs, 66

B

Bacterial remediation, 184
Barrier island, 62
Bathyal zone, 63
Bauxite, 108
Benthic ecosystem, 63–64
Benthic zone, 64
Beriberi, 96
Beta decay, 138
B-Horizon, 54
Bioaccumulation, 96
Bioassay, 178
Biodiversity
    climate changes and threats to, 200
    ecosystem biodiversity, 70, 206
    genetic, 69–70, 203
    protecting, 206–208
    species biodiversity, 70, 204–206
Biological communities, traits of, 58–60
Biological oxygen demand, 166
Biological populations, 58
Biomagnification, 96
Biomass energy, 145
Biomass pyramid, 66–67

Biomes
  aquatic, 63–64
  terrestrial, 61–62
Biotic factors, 58
Biotic potential, 78
Birth reduction pressures, 81
Bituminous coal, 134
Boiling water reactor, 140
Boreal forest, 62
Botanical pesticides, 100
Bottleneck effect, 203
Bureau of Land Management (BLM), 101, 114
Buy local, 123

C

Calcium ions, 168
California Water Project, 119–120
Calories, human nutritional requirement, 94
Canal, 112
Capping, 184
Captive breeding programs, 208
Carbamates, 100
Carbonate ions, in carbon cycle, 71
Carbon cycle, 70–71
Carbon dioxide, 155
  greenhouse effect and, 193
  increase in, and global warming, 194–195
  percent in atmosphere, 42
Carbon footprint, 201
Carbon monoxide, 155, 159
Carbon sequestration
  in carbon cycle, 70, 71
  forests and, 104
Carcinogens, 180–181
Carrying capacity, 80
  shifting, 89
CERCLA, 115
Channel, 64, 112
Channelization, 120
Chaparral, 62
Chemical erosion, 54
Chemical reactions, 24
Chemoautotrophs, 66

Chernobyl, 140
China, managing population growth, 89
Chloride, water pollution, 169
Chlorinated hydrocarbons, 100
Chlorofluorocarbons (CFCs), 157
C-Horizon, 54
Chromium, 108
Chromium salts, 159
Chronic toxicity, 182
Clay, as component of soil, 52
Clean Air Act, 125, 164, 202
Clean Water Act, 121, 173
Clean zone, 167
Clear-cutting, 104
Climate, 46–47
  climate shift, 68–69
  compared to weather, 46
  effects of climate changes, 199–201
  El Niño Southern Oscillation (ENSO), 47
  ocean currents and, 46–47
  reducing climate change, 202
Climate shift, 68–69
Coal, 71, 134, 135
Cold fronts, 46
Columbia River, 119
Commensalism, 60
Community, biological, 58–60
Competition, 60
Competition, interspecific and intraspecific, 60
Complexity
  food web, 66
  in species biodiversity, 69
Composting, 176
Compounds, 24
Comprehensive Environmental Response, Compensation, and Liability Act (CERCLA), 115, 185
Conclusion, in experimental design, 33
Conduction, 36
  global movement of energy and, 36

Confined aquifer, 49
Conservation
  of energy, 147–149
  land conservation options, 115
  mitigation, 115
  preservation, 115
  reclamation, 115
  remediation, 115
  restoration, 115
  soil, 97–98
  solid waste, 176–177
  of water, 120–121
  wetlands, 106
Conservation of matter, 24
Constructive boundaries, 38
Contour farming, 55, 97
Controls, in experimental design, 32
Convection, 36
  global movement of energy and, 36
Convergent evolution, 68
Convergent plate boundaries, 38–39
Copper, 108
Coral reef, 63
Core, Earths, 36–37
Coriolis effect, 44, 45, 46
  ocean circulation and, 50
Crime, as urban environmental problem, 110
Criteria pollutants, 153
Crude birth rate, 76
Crude death rate, 76
Crust, Earth's, 36–37

D

Data analysis, in experimental design, 32
Data tables
  analyzing, 31
  in experimental design, 32
Debt-for-nature swap, 207
Decomposition zone, 167
Deforestation
  factors contributing to, 104–105

global warming and, 194
as impact of agriculture, 99
rangelands and, 102
Demographic transitions, 86–87
Denitrification, 72
Density dependent factors, 80–81
Density independent factors, 81
Desert, 61
Desertification
as impact of agriculture, 99
rangelands and, 102
Developed countries, energy use by, 133
Developing countries, energy use by, 133
Dimensional analysis, 27
Dinoflagellate, 169
Disease, population growth and, 89–90
Disease clusters, 179
Disease virulence, 180
Dissolved oxygen, 166
Divergent evolution, 68
Divergent plate boundaries, 38
Dose-response curve, 179
Doubling time, 78
Dumps, 174–175

E

Earthquakes
focus of, 39
plate tectonics and, 39
Richter scale, 39
Earth systems, 35–55
atmosphere, 40–47
climate, 46–47
Earth's layers, 36–37
energy flows to and from Earth, 42–43
Geologic Time Scale, 39
global movement of matter and energy, 36
groundwater, 48–49
ocean circulation, 50
plate tectonics, 37–39
seasons, 43

soil, 51–55
water cycle, 47–48
weather, 45–46
E. coli, 166
Ecological niche, 58
Ecological role, 58
Economics
buying local, 123
external costs, 122–123
globalization, 123–124
Tragedy of the Commons, 121–122
Ecosystem biodiversity, 69–70, 206
protecting, 206–208
Ecosystems
aquatic biome, 63–64
biodiversity, 69–70
biological populations and communities, 58–60
biotic and abiotic factors, 58
carbon cycle, 70–71
climate shift and, 68–69
competition, 60
composition of, 58
energy flow in, 65
food chains, 66–67
keystone and indicator species of, 59
natural selection, 67–68
nitrogen cycle, 72–73
phosphorus cycle, 73
structure, 58–60
sulfur cycle, 73–74
symbiosis, 60
terrestrial biomes, 61–62
tolerance limits, 59
Edge effect, 207
Electrical energy, calculating, 130–131
El Niño Southern Oscillation (ENSO), 47, 199
Emergent diseases, 180
Emigration, 81
Endangered species, traits of, 204
Endangered Species Act, 125, 208

Energy
biomass energy, 145
biomass pyramid, 66–67
conservation of, 147–149
conversions of, 66
ecosystem energy flow, 65
electrical energy calculations, 130–131
energy flows to and from Earth, 42–43
energy pyramid, 66–67
First Law of Thermodynamics, 66, 128
fossil fuels, 134–137
geothermal energy, 146–147
global demand for, 136
global movement of, 36
history of human use of, 132–134
hydroelectric power, 144
kinetic, 128
nonrenewable sources of, 134–142
nuclear, 137–142
potential, 128
renewable sources of, 143–147
Second Law of Thermodynamics, 66, 129
solar energy, 143–144
specific heat calculations, 131–132
tidal, 146
total mechanical, 128
units and conversions of, 129–130
wind, 145
Energy conversions, 66
Energy crisis, 134
Environmental racism, 175
Environmental resistance, 79
Environmental risk analysis, 178–179
Erosion
gully, 54
as impact of agriculture, 99–100

rill, 54
sheet, 54
soil, 54
wetlands to help control, 106
Essential amino acids, 95
Estuaries, 64
Estuaries and Clean Waters Act, 115
Eutrophic lake, 64
Evolution
　convergent, 68
　divergent, 68
Exosphere, 42
Experimental design
　data analysis and conclusion, 32–33
　data and control, 32
　hypothesis, 32
Exponential growth, 78–79
Exponential notation, 25–26
External costs, 122–123

F

Famines, 96
Farming. *See also* Agriculture
　contour, 55, 97
　no-till, 55, 97
Fecundity, 76
Federal Aid Highway Act, 111
Federal Food, Drug and Cosmetic Act (FFDCA), 97
Federal Hazardous Substance Act, 185
Federal Highway System, 111
Federal Insecticide, Fungicide, and Rodenticide Act (FIFRA), 101
Federal land management, 113–114
Federal Wildlife Refuge system, 114
Fertility, 76
Fire management, 105, 207
First Law of Thermodynamics, 66, 128
Fish and Wildlife Act, 103, 115
Fishing, 102–103

Fission, 138
Floods, wetlands as flood control, 106
Flue gas desulfurization, 155
Fluidized bed combustion, 155
Food
　basic requirements for humans, 94–95
　effects of population growth on, 89–90
　famines, 96
　genetically modified organisms (GMOs), 98
　sustainable agricultural methods, 98–99
　toxins in, 96
　undernourished *vs.* malnourished, 95–96
Food chains, 66–67
　biomass pyramid, 66–67
Food webs, 66
Fore reef, 63
Forests
　in carbon cycle, 71
　carbon sequestration and, 104
　clear-cutting, 104
　factors in deforestation, 104–105
　fire management, 105
　national, 114
　old-growth, 104
　sustainable forestry, 105
　Swidden agriculture, 104–105
　tree plantations, 104
Forest Service, 114
Fossil fuels
　in carbon cycle, 71
　formation, extraction and combustion of, 134–137
　formation of, 71
　pros and cons of using, 137
　in sulfur cycle, 73
　types of, 134–135
　world reserves of, 135–136
Founder effect, 69, 203
Fracking, 107, 135
Freshwater, 48

Fugitive emissions, 152
Fusion, 138

G

Gap analysis, 207
Gene expression, 69
Gene pool, 69, 203
Genes, 69, 203
Genetically modified organisms (GMOs), 98
Genetic assimilation, 203
Genetic biodiversity, 69–70, 203
Genetic isolation, 69
Genetics
　genetic biodiversity, 69–70
　natural selection and, 67–68
Geologic Time Scale, 39
Geothermal energy, 146–147
Giardia, 166
Gill netting, 102
Globalization, 123–124
Global warming, 192–202
　Albedo effect, 194
　carbon dioxide levels, 194–195
　carbon footprint, 201
　climate changes, 199–201
　decreased ocean pH, 198
　deforestation, 194
　evidence of, 194–198
　greenhouse effect, 192–193
　ice core data, 196
　laws relating to, 202
　mechanisms that accelerate, 193–194
　methane sequestration, 193–194
　polar ice melting, 197
　reducing climate change, 202
　sea levels rising, 197
　sea temperature rising, 197–198
　temperature changes, 194–195
Goiter, 95

Gold, 108
Graphite reactor, 140
Graphs, analyzing, 29–30
Grasshopper effect, 164
Grasslands, 62
Gravel, as component of soil, 52
Greenhouse effect, 163, 192–193
Greenhouse gases, 193
  carbon footprint, 201
  reducing climate change, 202
Green Revolution, 98
Gross primary productivity (GPP), 65
Groundwater, 48–49
Groundwater pollution, 171
Growth rate, calculating, 77
Gully erosion, 54

H

Habitat destruction, population growth and, 90
Hadal zone, 63
Half-life, 138–139
Hard water, 168
Hazardous chemicals in environment
  cleanup of contaminated sites, 184
  natural remediation of, 183–184
  types of, 183
Hazardous Materials Transportation Act, 185
Heat domes, 164
Heat island effect, 110
Heavy water reactor, 140
Helium, percent in atmosphere, 42
Heterotrophs, 66
Highway system, 111
Histamine, 181
Hooks, fishing with, 103
Hoover Dam, 118
Human populations
  age-structure diagrams, 85–86
  demographic transitions, 86–87
  distribution of, 85
  dynamics of, 84–86
  managing population size, 88–89
  strategies for sustainability, 88–89
Humus, as component of soil, 53
Hunger, population growth and, 89–90
Hunter-gatherers, energy use by, 132
Hybrid cars, 148
Hydroelectric dam and reservoir projects, 118–119
Hydroelectric power, 144
Hydrogen, percent in atmosphere, 42
Hydrogen fuel cells, 144
Hyperthyroidism, 95
Hypothesis, in experimental design, 32

I

Ice core data, 196
Igneous rocks, 51, 52
Immigration, 81
Incineration, 175, 184
India, managing population growth, 89
Indicator species, 59
Indoor air pollution, 159–160
Industrial Revolution, energy use during, 133
Industrial water use, 117
Inorganic pesticides, 100
Integrate pest management systems, 101
Interspecific competition, 60
Interspecific factors, 80–81
Intertidal ecosystem, 64
Intraspecific competition, 60
Intraspecific factors, 81
Inversions, 164
Ions, 24
Iron, 108
water pollution, 169
Irrigation methods, 97
Isotopes, 23

J

Jet stream, 45
Joule, 129

K

Keystone species, 59
Kinetic energy, 128
Krakatau Island, 40
Kwashiorkor, 95
Kyoto Protocol, 124, 165, 202

L

Lacey Act, 125
Lake, ecosystem, 64
Landfills, 174–175
Land reclamation, 207
Land use
  agriculture, 94–100
  fishing, 102–103
  forests, 104–105
  land conservation options, 115
  land reclamation, 207
  land use mitigation, 208
  mining, 107–109
  pest control, 100–101
  public and federal lands, 113–114
  rangelands, 101–102
  transportation infrastructure, 111–113
  urban land development, 109–111
  wetlands, 105–106
Laws of Thermodynamics, 66, 128–129
LD_{50}, 179
Lead, 108
  from air pollution, 157, 159
LEED certification, 111
  defined, 186
  goals of, 186

requirements of, 186–188
Life expectancy, 76
Lifespan, 76
Lignite coal, 134
Limestone, in carbon cycle, 71
Limnetic zone, 64
Lithosphere, 36
Littoral zone, 64
Loam, as component of soil, 53
Lock and dam projects, 118–119
Logistic growth, 79
Lower mantle, 36
Low-flow utilities, 120
Lungs, air pollution as irritant, 161

M

Magnuson-Stevenson Fisheries Management and Conservation Act, 103
Manganese, 108
Mangrove swamp, 64
Mantle, Earth's, 36–37
Maps, analyzing, 28–29
Marasmus, 95
Marianas Trench, 39
Math skills
  dimensional analysis, 27
  exponential notation, 25–26
  percentages, 26–27
  problem solving with units, 27–28
Matter
  conservation of, 24
  global movement of, 36
  principles of, 23–24
  states of, 25
Mature communities, 69
Mechanical weathering, 51
Mercury
  from air pollution, 157
  water pollution, 168
Mesosphere, 42
Metals, as air pollution, 157
Metamorphic rock, 51, 52
Methane, percent in atmosphere, 42

Methane sequestration, 193–194
Mid-Atlantic ridge, 38
Minamata disease, 168
Mineral rights, 108
Minerals
  formation of, 107
  for human nutrition, 95
Mining, 107–109
  methods of, 107
  worldwide mineral reserves, 108
Mitigation, land, 115
Mono Lake, 120
Montreal Protocol, 124, 165, 192
Morbidity
  defined, 76
  environmental hazards and, 178
Mortality
  defined, 76
  environmental hazards and, 178
Mutagens, 180–181
Mutations, 68
Mutualism, 60

N

Natality, 76
National Environmental Policy Act (NEPA), 124
National parks and forests, 114
National Park Service Act, 125
Natural gas, 135
Natural growth rate, 76
Natural selection
  defined, 67
  elements of, 68
  genetic basis of, 67–68
Nazca plate, 38
Neolithic era, energy use in, 132
Neon, percent in atmosphere, 42
Neritic zone, 63
Net primary productivity (NPP), 65

Neurotoxins, 181
Neutron capture, 138
Neutrons, 23
Nickel, 108
Nitrates, 154
Nitrification, 72
Nitrites, 154
Nitrogen, percent in atmosphere, 42
Nitrogen cycle, 72–73
Nitrogen fixation, 72
Nitrogen oxides, 154
Noise Control Act, 165
Noise pollution, 165
Noncriteria pollutants, 153
Nonrenewable energy sources, 134–142
No-till farming, 55, 97
Nuclear energy, 137–142
  defined, 137
  half-life, 138–139
  measuring radiation doses, 141
  nuclear reactors, 139–140
  pros and cons of, 142
  radiation and human health, 141
  radioactive waste, 142
  terms for, 137–138
Nuclear reactions, 137
  rates of, 138
Nuclear reactors, 139–140
Nutrition
  basic requirements for humans, 94–95
  toxins in food, 96
  undernourished vs. malnourished, 95–96

O

Occupational Safety and Health Act (OSHA), 124
Ocean Dumping Ban Act, 121
Oceans, 198
  decreased pH of, 198
  ecosystems, 63–64
  El Niño Southern Oscillation (ENSO), 47

global warming and sea
  levels/temperature,
  197–198
ocean circulation, 50
thermohaline currents, 50
tides, 50
weather and, 46–47
O-Horizon, 53
Oil, 134–135, 136
  water pollution, 169
Old-growth forests, 104
Oligotrophic lake, 64
Oncogene, 180–181
Open ocean ecosystem, 63–64
Organism, 58
Organophosphates, 100
Overfishing, 103
Overgrazing, 102
Oxygen
  creation of, in atmosphere, 40
  oxygen-demanding waste, 166–167
  percent in atmosphere, 42
Oxygen sag curve, 166–167
Ozone
  causes and effects of depletion, 191–192
  depletion of, 163–164
  formation of, 190
  percent in atmosphere, 42
  strategies for reducing depletion, 192
  ultraviolet radiation, 190

P

Paralytic shellfish poisoning, 169
Parasitism, 60
Park Service Act, 115
Particulate matter, 156
Passive solar energy, 143
Pathogen, 180
Pelagic ecosystem, 63–64
Pellagra, 96
Percentages, problem solving with, 26–27
Permeability, of soil, 53

Pesticides
  costs and benefits of, 101
  in food, 96
  integrate pest management systems, 101
  types of, 100
pH
  decrease in ocean pH, 198
  defined, 24
  of soil, 53
Phosphorus cycle, 73
Photic zone, 63
Photoautotrophs, 66
Photochemical oxidants, 158
Photosynthesis
  in carbon cycle, 70, 71
  gross primary productivity, 65
  net primary productivity, 65
Photovoltaic panels, 144
Pioneer species, 68–69
Plate tectonics, 37–39
Polar ice, in carbon cycle, 71
Polar ice melting, 197
Pollution
  acid rain, 154
  air, 152–165
  anthropogenic sources, 152
  criteria pollutants, 153
  environmental risk analysis, 178–179
  hazardous chemicals in environment, 183–184
  hazards to human health from, 177–182
  laws relating to, 185
  LEED certification, 185–188
  noise, 165
  noncriteria pollutants, 153
  Precautionary Principle, 177
  solid waste, 173–177
  water, 166–173
  weather and, 164
Ponds, ecosystem, 64
Populations
  biological, 58
  carrying capacity, 80, 89

defined, 58
demographic transitions, 86–87
effects of population growth, 89–90
factors affecting size of, 80–83
growth rate, 77–80
human population dynamics, 84–86
key terms of, 76
managing population size, 88–89
reduce fragmentation of, 206–207
strategies for sustainability, 88–89
survivorship curve, 83
Porosity, of soil, 53
Potential energy, 128
Power, calculating, 130
Precautionary Principle, 177
Preservation, land, 115
Pressure, weather and, 45
Pressurized water reactor, 140
Primary productivity, 65
Primary succession, 68–69
Problem solving, with units, 27–28
Producer, 66
Productivity
  in ecosystems, 65
  primary, 65
  secondary, 65
Profundal zone, 64
Pronatalist pressure, 81
Protons, 23
Public lands, 113–114
Purse seining, 102

R

Radiation
  causes/effects of, by dose, 141
  defined, 36, 141
  global movement of energy and, 36
  measuring doses of, 141

radioactive waste, 142
ultraviolet radiation, 190
Radon, 159
Rain, acid, 134
Rangeland management, 101–102, 207
Reclamation, land, 115
Recovery zone, 167
Recycling, 148
solid waste, 175–176
Reduce, reuse, recycle, 175–176, 184
Reef crest, 63
Reef lagoon, 63
Remediation, land, 115
Renewable energy sources, 143–147
biomass energy, 145
geothermal energy, 146–147
hydroelectric power, 144
solar energy, 143–144
tidal energy, 146
wind energy, 145
Reproductive strategies, 82
Residential water use, 117
Resistance, 180
natural selection, 68
Respiration, in carbon cycle, 70, 71
Restoration, land, 115
Richter scale, 39
Rickets, 96
Rill erosion, 54
Ring of Fire, 40
Riparian zone, 64
Rock cycle, 51, 52
chemical weathering, 51
Rocks
igneous, 51, 52
mechanical weathering, 51
metamorphic, 51, 52
sedimentary, 51, 52
Rocky beach, 64

S

Safe Drinking Water Act, 121, 173
Salinity, of soil, 53
Salinization, as impact of agriculture, 100
Sand, as component of soil, 52
Sandy beach, 64
Savannah, 62
Schistosomiasis, 166
Scientific data, analyzing
data tables, 31
graphs, 29–30
maps, 28–29
Scurvy, 96
Sea floor spreading, 38
Seasons, 43
Secondary productivity, 65
Secondary succession, 69
Second Law of Thermodynamics, 66, 129
Sedimentary rocks, 51, 52
Septic zone, 167
Sewage treatment, 171–172
Sheet erosion, 54
Silt, as component of soil, 52
Silver, 108
Soil
composition of, 52–53
conservation methods, 55
erosion of, 54
formation of, 52
nutrient depletion as impact of agriculture, 100
physical and chemical properties of, 53–54
rock cycle, 51–52
soil conservation, 97–98
soil horizons, 53–54
types of rock, 51
Soil Conservation Act, 100
Soil horizons, 53–54
Solar energy, 143–144
Solid waste, 173–177
dumps and landfills, 174–175
incineration, 175
reduce, reuse, recycle, 175–176
toxic colonialism, 175
types of, 173–174
waste disposal, 174–175
South American plate, 38

Species
competition, 60
symbiosis, 60
Species biodiversity, 70, 204–206
benefits of, 204
minimum viable populations, 205
protecting, 206–208
threats to, 204–205
traits of endangered species, 204
Specific heat, calculations, 131–132
States of matter, concept of, 25
Stratosphere, 41
Stratospheric ozone. See Ozone
Streams, ecosystem, 64
Subsurface mining, 107
Suburban sprawl, 110
Sulfate ions, 73
Sulfide ions, 74
Sulfur, oxides of, 154–155
Sulfur cycle, 73–74
Sulfur dioxide, 154–155
Sulfur oxides, 73
Sun, energy flows to and from Earth, 42–43
Supratidal island, 62
Surface mining, 107
Surface Mining Control and Reclamation Act (SMCRA), 109, 115, 207
Survivorship, 76
Survivorship curve, 83
Suspended particles, 167–168
Sustainability
of fishing, 103
forestry, 105
strategies for, 88–89
sustainable agricultural methods, 98–99
transportation and, 112–113
urban land development, 110–111
Swidden agriculture, 104–105
Symbiosis, 60
commensalism, 60

mutualism, 60
parasitism, 60

T
Tables, data, analyzing, 31
Taiga, 62
Temperate deciduous forest, 62
Temperate rainforest, 62
Temperature
    global temperature changes, 194–195
    weather and, 45
Teratogens, 181
Terracing, 55, 97
Terrestrial biomes, 61–62
Thermal pollution, 170
Thermohaline currents, ocean circulation and, 50
Thermosphere, 42
Three-Gorges Dam, 119
Threshold level of toxicity, 179
Tidal energy, 146
Tides, ocean circulation and, 50
Tin, 108
Tolerance limits, 59
Total growth rate, 76
Total mechanical energy, 128
Toxic colonialism, 175
Toxic materials
    factors affecting toxicity, 181–182
    mechanisms of, 180–181
Toxins, in food, 96
Trade winds, 45
    ocean circulation and, 50
Traffic, as urban environmental problem, 109
Tragedy of the Commons, 121–122
Transform boundaries, 39
Transportation
    channels and canals, 112
    Federal Highway System, 111
    sustainable, 112–113
Traps, fishing, 103
Trawling, 103
Tree plantations, 104
Trombe walls, 143
Trophic level, 66
Tropical cyclones, 46
Tropical rainforest, 62
Troposphere, 41
Tumor suppressor gene, 181
Tundra, 61

U
Ultraviolet radiation, 190
United States, managing population growth, 89
Units, problem solving with, 27–28
Upper mantle, 36
Urban land development, 109–111
    environmental problems of, 109–110
    sustainable, 110–111
    trends in, 109

V
Vaccines, 180
Vectors, 180
Vitamins, 95
Volatile organic compounds (VOCs), 155–156, 159
Volcanism, 40
Volcanoes, plate tectonics and, 38

W
Warm fronts, 46
Waste disposal, 174–175
Waste stream, 173
Water
    as cause of erosion, 54
    confined aquifer, 49
    freshwater, 48
    groundwater, 48–49
    moisture and weather, 45
    municipal purification, 171
    ocean circulation, 50
    water compartments, 48, 49
    water cycle, 47–48
    well water, 49
    wetlands purification, 106
Water compartments, 48, 49, 116–117
Water cycle, 47–48
Water pollution, 166–173
    dissolved ions, 168–169
    groundwater pollution, 171
    inorganic waste, 167–169
    laws relating to, 173
    maintaining water quality, 171–172
    nonpoint sources, 166
    oxygen-demanding waste, 166–167
    pathogens, 166
    point sources of, 166
    sewage treatment, 171–172
    suspended particles, 167–168
    thermal pollution, 170
    toxic organic waste, 169
    types and sources of, 166–170
    as urban environmental problem, 110
    water purification, 171
Water Quality Act, 121
Watershed management, 120
Water use
    agricultural, 117
    channelization, 120
    conservation methods, 120–121
    fishing, 102–103
    industrial, 117
    irrigation methods, 97
    lock and dam projects, 118–119
    managing water resources, 118–119
    residential, 117
    trends in, 116–117
    water diversion projects, 119–120
    wetlands, 105–106
Weather, 45–46
    cold and warm fronts, 46
    compared to climate, 46

defined, 46
pollution and, 164
temperature, pressure and moistures affect, 45
tropical cyclones, 46
Weather fronts, 46
Wegener, Alfred, 37
Wells, mineral extraction, 107
Well water, 49
Wetlands
conservation of, 106
importance of, 105–106
protecting, 207
Wet scrubbing, 155
Wildlife refuges, 114
Windbreak trees, 55, 97
Wind energy, 145
Winds
as cause of erosion, 54
trade winds, 45

Z

Zinc, 108